Trials of a Darwin Doubter

Endorsements

I can count on both of my well-designed thumbs the novels I have read about the creation-evolution debate. Dr. Bergman, well known for his many scholarly books on the subject, has mutated into writing fact-based fiction. If one were to think the novel would be the story of a creationist who prevails over extreme persecution and maintains the high road until victory is achieved, the reader would be surprised. A mixture of science and suspense, *The Trials of a Darwin Doubter* will open eyes and (hopefully) minds as well.

—Karl Priest, MA Educator.

Dr. Jerry Bergman, with many scientific degrees (including doctorates), has been grappling with evolutionary indoctrination in science for decades. I was particularly attracted to his new approach. He documents the statement in Chapter 2: "natural selection is the opposite of evolution. It reduces variety and eliminates many life-forms. It does not produce new lifeforms, but in the end only causes the extinction of many different varieties of life." In Chapter 9, the response to a question, "How do you know that evolution is fact?" was naïve: "Because all credible scientists accept it. That's why." Dr. Bergman is, in all his writings, a voice crying out, the "Evolution Emperor is Naked!," but responses so often seem to be like blind, cave-dwelling fish: having lost vision, constantly bumping into the corrective walls of truth.

—Paul Humber, College Educator

Trials of a Darwin Doubter

Inspired by a True Story

SECOND EDITION

By Jerry Bergman, PhD

Creation Summit Publishers
LINDENHURST, ILLINOIS

Published by
Creation Summit Publishers, Lindhurst, Illinois
https://creationsummit.com/

Printed in the United States of America

Distributed by
Amazon.com

Because of the dynamic nature of the internet, any web addresses or links contained in this book may have changed since publication and may no longer be valid.

Views expressed in this work are solely those of the author and do not necessarily reflect the views of the publisher, and the publisher hereby disclaims any responsibility for them. Furthermore, the names of persons and institutions were changed to protect the actual persons and institutions. This includes Cornell and Harvard universities, neither of which were, in most cases, part of the actual events in this book.

Bergman, Jerry, 1946–
Trials of a Darwin doubter. Second edition.

ISBN: 979-8-9950778-3-1

Contents

Preface

This story is a composite based on the published, well-documented case histories in my books *Slaughter of the Dissidents: The Shocking Truth About Killing the Careers of Darwin Doubters,*[1] *Silencing the Darwin Skeptics,*[2] *Censoring the Darwin Skeptics,*[3] *Suppressing the Darwin Skeptics,*[4] and those cases that are part of new manuscripts in this series, now in preparation. Many of the cases in these publications have numerous elements in common; thus, this composite story is typical of many such cases. Much of the dialogue was taken from manuscripts of actual court cases or conversations. Only the last chapter was not directly based on actual events. Some names were changed, including those of the universities involved, to protect the privacy of certain people.

If one has read a single case involving suppression of Darwin doubters, the reader will note that many major similarities exist with this case history. This account will allow the reader to understand the antagonism against those who have valid scientific questions about the validity of orthodox Darwinism. Most of the obscene or scatological profanity sometimes used by Darwinists has been expunged. Importantly, Cornell was not the university where the story took place but was used as a setting for several reasons. Most of the names of the professors involved in the dialogues were also changed to protect the innocent as well as the not-so-innocent.

---

[1] 2008. Revised version, 2012. Southworth, WA: Leafcutter Press, 482 pp.

[2] Southworth, WA: Leafcutter Press, 2016, 396 pp.

[3] Southworth, WA: Leafcutter Press, 2018, 400 pp.

[4] In preparation. Southworth, WA: Leafcutter Press.

Acknowledgements

Among the many persons contributing to this work are Bryce Gaudian, Marilyn Dauer, Karl Priest, Kirk Toth, David V. Bassett, and Roger Lovell.

Chapter 1: My First Christmas

Early one Christmas morning, I was, to say the least, an excited ten-year-old looking forward to the big day. I had wanted a good, high-powered microscope for several years and, as I opened the set of presents from my parents, I soon had a microscope in front of me. A real microscope! It was a used medical microscope, but a real professional Bausch and Lomb medical microscope. It had 10, 43, and 100 power Karl Zeiss objectives, and a ten-power eyepiece, achieving a total of 100, 430, and 1,000 power. I was in heaven!

Yes, even as nonobservant Jews, we Radinows celebrated Christmas. We had assimilated almost imperceptibly into American upper-middle-class culture. Not much difference existed between our neighbors and us. I was, though, very much aware of being Jewish, especially since the Holocaust had occurred only a few decades ago, and we lost most of our relatives who were still back in our homeland when the war broke out. We had retained a love of education, books, and learning common to many Jews. This trait would, in time, get me into big trouble later in life. But more on that later.

As the months went by, I placed onto my microscope's mechanical stage everything I could find that fit in between the microscope's objectives and its stage. Pond water, pollen, yeast, flower petals, plant stems, different kinds of paper, insect wings, animal hair, money, a diamond ring, thin rock chips of every type, soil, and anything living, or once living, that I could locate on the small farm where we then lived. I knew since sixth grade that I wanted to be a scientist, and my microscope explorations helped to firm up that conviction. Next Christmas came a good erector set, then a chemistry set, a radioactive lab, and the next year a rocket kit. I was well on my way to becoming a scientist!

My father purchased the microscope at the University of Michigan medical bookstore in Ann Arbor, where he was a chemistry professor. We would often go to his lab on the weekends, and I fondly remember the two large dark-stained oak cabinets packed with chemicals in the storage room and the large slate blackboards covering one wall of his chemistry classroom. He would often bring home small bottles of various chemicals for me to experiment with.

And, yes, I made black powder and other types of explosives, as did my friends. You could buy all of the ingredients at a drug store then. I used potassium nitrate, which was popularly called saltpeter, carbon in the form of powdered charcoal, and sulfur (both were sold under their chemical names). I learned to chemically "fuse" the gunpowder by making a water paste mixture, shaking it vigorously, pouring it on a cookie sheet, and, lastly, allowing the mixture to dry for several days. Next, I packed the powder in a five-inch-long pipe, capped the ends, drilled a small hole at one end, and placed a fuse into the hole, making a working bomb.

Once I completed assembling the first bomb, I went to an empty field to test it. I lit the fuse, ran like crazy, and waited. And then I waited some more. Disappointed, I began walking back, and within a few minutes heard a terrific explosion. Back at the site of the pipe bomb, after searching for close to twenty minutes, I was not able to even find any fragments. The pipe bomb must have shattered and scattered far and wide. Later, I was almost injured once, and a close friend was blinded in both eyes while experimenting with explosives. After that, I became much more careful. When young, one does some very foolish things.

In ninth grade, I was placed in advanced classes and took chemistry, biology, civics, English, and algebra. One issue I would later study in detail surfaced early in school. I always liked to bring up controversial topics because I quickly became bored with dull lectures—most of what was covered in my science class I already knew—so I often got some exhilarating discussions going to enliven the class. The students liked the excitement, as—*usually*—did my teachers.

One day, out of curiosity, I brought up the creation-evolution controversy in my biology class, and my teacher went ballistic. As a result of my questions, the entire class was treated to a powerful twenty-minute spontaneous lecture on the overwhelming evidence for Darwinism—the idea that from a single cell evolved all life-forms existing today. The evidence presented, which I vividly remember to this day, included the human tail. If God created humans, why would we have a tail like a monkey when we were an embryo? And why would humans have over one hundred vestigial organs (which we were then taught were degenerate, or even useless, organs), such as the appendix, wisdom teeth, and tonsils, all of which were once very useful in our evolutionary ancestors?

Our teacher also argued that ontogeny recapitulates phylogeny, meaning that as we develop in the womb, we pass through all of our major past evolutionary stages. All humans began as one cell, then became a multicellular organism like slime mold, then a longer, more complex worm-like organism. Next, we evolved into a fish, then into an amphibian, next a reptile, a mammal, a primate, and, lastly, into a human. All of these evolutionary stages were passed through by humans during the nine months we were in our mother's womb.

This story, we were all told authoritatively, was all proven beyond question by science, namely Haeckel's convincing embryo pictures accurately drawn from real life that were published in our biology book, which was written by several leading biology professors. Haeckel's theory was also proven in our young minds by the fact that human embryos had a really small tail and gills just like fish, documenting our ape and fish ancestry.

Our teacher then discussed the human fossil record, including Piltdown Man, Neanderthal Man, and Java Man, all of which, he stressed, had proven beyond a doubt that humans evolved from some apelike common ancestor. One of the best examples in our outdated biology textbook, Piltdown Man, looked very convincing to me, even though I found out later that it had been proven to be a hoax in 1953.

Another proof of evolution included in our biology text was *homology,* the idea that structural similarities, such as in the bones of various animals, proved that we evolved from some common ancestor. If this view were not true, why would our forearm bone set look so much like those of other animals, including dogs, lions, birds, and even bats?

The poor design claim was another argument our teacher used to discredit creation. He explained that if God created us, evidence of poor design would not exist in the human body, but that many examples do exist. The examples he gave included the so-called backward retina—the light-sensitive rods and cones face *away* from the light, not towards it as one would expect. Another was the commonplace back and knee problems, considered proof of our evolution from quadrupedal apes, because these structures originally evolved in the early tetrapods from which we evolved. These were a few of the many examples of poor design showered on us that fateful day. Our teacher concluded that all of this scientific evidence proved the "from atoms to humans" evolution beyond a doubt by the natural selection of changes to our DNA, called mutations.

He also hammered home his conclusion that he believed in God, but God had better things to do with His time than create us. Besides, if God created us, He is to blame for all of our many imperfections. Rather, he argued, we evolved by natural selection, and those who denied this fact were ignorant, if not just plain stupid.

He was a good teacher, was very well-liked by his students, and made his point very effectively. We got the message loud and clear. This lecture, and so many others like it, moved me toward the Darwinism side of the controversy, but I still had some doubts. It was just my personality. I had doubts about a lot of things, always did, and probably always will. My parents, who were both convinced evolutionists, called me a "Doubting Thomas" while at the same time encouraging my skeptical attitude. After I completed eighth grade, we moved to Brooklyn, New York, because my father became chairman of the chemistry department at Columbia University.

I loaded up on science classes at the prestigious Brooklyn Stuyvesant High School, which I then attended. Evolution was hammered into our heads in almost every course, even in our math classes. It seemed that every time I asked *why* some biological structure was designed in some particular way I was told "It evolved that way." To me, this was not an answer, and it appeared to my young mind that evolution was more of a science stopper than a means of gaining knowledge and insight into the reasons for something. It reminded me of a common reason my mother gave as to why I was not allowed to do something (like stay outside until midnight), namely the single word "Because."

I once asked my high school biology teacher why so many birds, often males, were dressed in bright colors, but the color of most birds was an ugly, dull black or gray. I was told only that "they just evolved that way due to the natural selection of mutations."

"But why do so many bright colors exist on some birds?" I asked again.

Our biology teacher, Dr. Riekeman, answered, "Because the bright colors evolved to attract mates. The brighter birds attracted more mates; thus, the males were more likely to reproduce."

"But how did they attract mates until they evolved their bright colors?" I asked, always curious about everything.

I probably asked more questions than any other student in the class. Many of my fellow students, most of whom were the sons and daughters of successful professionals, just sat there, memorized a bunch of facts, and earned their As. Most of the kids in my class were very bright, and most had photographic memories, but they were not always very creative or inquisitive. At least that was my opinion.

"Well," Dr. Riekeman answered, "birds with a little brighter color attracted more mates than those with less bright colors. Sexual selection slowly caused the birds' color to become brighter and brighter, until today we have bright yellow, red, and blue birds," he answered confidently.

"But the bright colors will also attract predators," I continued, "and the brighter the colors, the more visible they are to predators. Brightly colored birds attract mates, which is good, but also attract enemies, which is bad."

Our teacher answered confidently, "That is why so many birds do not sport bright colors. Those birds void of bright colors are less likely to be consumed by predators and thus are more likely to be favorably selected by natural selection. For this reason, bright colors are selected against, causing the evolution of less bright colors!"

He then added, "Actually, this is why most birds are black, gray, or close to gray in color. Not very many birds have bright colors; we just remember them better because they stand out from the rest."

"Isn't it a contradiction," I added, "to conclude that natural selection simultaneously favors both brightly colored and less brightly colored feathered birds?"

"Well, that's how they evolved by natural selection. Natural selection sometimes works in mysterious ways," he noted with a smile, proud of his retort.

I added, "It seems to me that evolution tries to explain everything by the survival advantages argument, especially mating and avoiding predators. Isn't that very simplistic?"

"Well, ultimately all of biology, all life, can be explained by mutations occurring and the survival of the fittest mutants, as Darwin taught," he answered authoritatively.

"All biology?" I asked.

"Yes, all of biology," he replied. "And all of life, actually," gesturing with his hands to emphasize the point.

Dr. Riekeman then answered my two-word question with another one of his short lectures that he was well-known for, "In short, evolution stresses that all life produces more offspring than can be supported on our Earth. Each maple tree produces many thousands of seeds, and only a very few seeds take root. Seeds that have larger sails blow farther away in the wind and thus are more likely to take root. Those seeds that do take root are more evolved, even if just slightly, thus better adapted than the others. Even fewer of the seeds that take root grow to become mature trees, again due to the survival of the fittest law."

"But," I asked, "isn't chance a major factor? Those seeds that happen to land on fertile ground and, by chance, are not eaten by birds, are far more likely to sprout. Also, we have maple seeds entombed in what scientists claim are 50-thousand-year-old amber, and these fossils, as far as we can tell, are identical to those existing today."

"Let me give a better example," Dr. Riekeman added, changing the subject. "Sea turtles produce a catch of eggs that they lay on land in shallow holes, which are often robbed by snakes and scavengers. Of those eggs that hatch, the young begin their trek to the sea; many are consumed by carnivores. Even those turtles that make it to the sea are often consumed by carnivores before they can reproduce. Out of every one hundred eggs laid, only four may survive to become mature turtles.

I said, "From what you explained, it seems to me that chance is critical to who survives, at least with sea turtles. The eggs that happen to be at the bottom of the sea turtle nest would be less likely to be eaten, at least at first. Or those at the top may be less likely to be eaten, depending on when the predators arrive. Survival of the fittest? A two or three-inch-long baby sea turtle cannot outrun a dog."

Riekeman said, "No, but in the end, only the fittest will survive, a process called natural selection, which was Darwin's great contribution to science. After many generations of evolution, turtles will eventually evolve into some other animal, a more fit animal. That's how turtles got here in the first place."

"What were turtles before they evolved into turtles?" I asked.

"Scientists do not know the answer to that question yet. All of the fossil turtles so far discovered are clearly modern turtles, although some are extinct turtle types, but all are clearly turtles. I am sure that we will eventually find the evolutionary ancestors of turtles in the fossil record. We know that they must have evolved from some shell-less reptile or amphibian. We just need to find the evidence, and we will continue to spend millions of man-hours and dollars until we do."

"But what if we never find a shell-less proto turtle?" I asked innocently.

"Then, since we know the fossils of animals evolving into turtles existed, scientists will have to develop a theory about why evidence for them has never been found. They evolved, and will continue to evolve in the future, and forever. That is a fact!"

"And just what will sea turtles evolve into?" I pressed him.

"Who knows? Evolution may cause sea turtles to lose their shell to help them become more streamlined, allowing them to swim much faster and escape their predators more effectively."

I continued, "But our textbook claims that turtles evolved a shell to protect themselves from predators, a strategy that works very well, at least for adult turtles, so why would they lose their shell?"

I then added while shifting in my seat, "Our textbook shows that 150-million-year-old turtles are just turtles, so if they have not evolved in 150-million years, why do we think they will evolve into something else in the next few million years?"

"Because all life continually evolves to become better adapted to its local environment," he answered confidently.

"If they evolved a shell, why lose it?"

"I told you, so that sea turtles can swim faster and more effectively escape predators as your textbook states," he answered, thinking his response demolished my argument.

"Then why did they evolve a shell in the first place? Why not just evolve the equipment to swim faster rather than evolve a shell, which would not help them much in escaping predators? Nor would it help to protect them until their shell was fairly complete, at least complete enough to protect them." I protested.

"You ask far too many questions," Dr. Riekeman answered, somewhat exasperated now.

I then noted pensively that, "The baby turtles were heading en masse to the ocean to serve as dinner for the birds along the way. The turtles that survive their journey to the ocean are, by far, most often not the most fit, just the most fortunate. And the lion that picks out the newborn calf for a meal rather than the strong bull makes a rational selection. The calf is not less fit than the bull, just young and in the wrong place at the wrong time."

I thought about these evolutionists' claims for a while, and from these examples, my thought was that the facts often show not survival of the fittest, but survival of the luckiest. The turtles that survive are most often not the fittest turtles, and those that don't survive are often not less fit, but only the ones that are in the wrong place at the wrong time. What if it just happened that no predators were around where the turtles were one day? Then all of the turtles would make it to the water.

"Well, we need to move on so we can cover the required class material," Dr. Riekeman shot back, interrupting my train of thought and changing the subject.

I also thought of many other examples that did not fit his story, such as where I had read that a spider fossil, dated by evolutionists to be 165 million years old, was found recently, and it looks exactly like a modern spider. It even had spinnerets just like modern spiders do.

Also, I thought about humans in most parts of the world, or animals that have low reproduction rates, such as gorillas, which are now threatened with extinction. I knew that animals with few offspring were more nurturing and protective, thus more of their offspring survived. It would be better if they had more offspring, and each one needed less parental attention. Of course, Dr. Riekeman never mentioned any of these big problems in his evolution theory.

Dr. Riekeman continued, "All life exists in an enormous variety, and natural selection just prunes the existing variety."

This prompted me to ask, "But where did all of that variety originally come from?"

This question really frustrated Dr. Riekeman, and it showed up in his voice. After again stating, "You ask too many questions," he realized, judging by the class's reaction, that this response was not appreciated, so he responded by noting, "Variety is just a fact of nature. Variety exists everywhere, in geology, in rocks, in the ninety-two natural elements, and in all life from bacteria to plants, animals, and even snowflakes and humans. Just look around this class, every one of you is different, and has different hair, skin, and eye colors; different heights, different body builds, different noses, and mouth shapes," (a comment that elicited some chuckles from the class—rare chuckles, I might add).

"OK, we get the picture," a fellow student David added, somewhat bored at this point.

"But," I protested, "to note that a lot of variety exists in nature still does not answer the question where it all came from! It only makes the source of this variety a much more important problem."

Dr. Riekeman, still registering frustration in his voice, came back with, "For all life, the ultimate source of all variety is mutations, which are mistakes in DNA copying, or damage to the DNA caused by cosmic rays or chemical mutagens. That's where all of the variety in life ultimately comes from."

"Are mutations the *source* of the amazing and enormous variety of life? Certainly, it is not due to DNA damage from new mutations," I asked Dr. Riekeman.

To this inquiry, Dr. Riekeman responded, "That's a very good question. Well, we will have to work on the problem of the origin of variety in nature. Scientists do not have all the answers, not yet at least, but they will! I have a lot of faith that science will answer those questions someday."

I responded, "I just read that more than a single mutation is an insurmountable barrier to evolution. It is a major problem for Darwinism since the probability at each nucleotide locus is 1 out of 100 million per generation. To get two specific mutations that function together to further evolution, the rarity is squared into trillions. Thus, it will only very rarely, if ever, happen. Three coordinated functional mutations simply won't happen in known geologic time for any species."

At this point, Dr. Riekeman, after glancing at the clock, beamed, "Class is over. For tomorrow, read chapters 7 and 8 that cover the overwhelming proof of evolution. See you tomorrow."

In another class, later that week, my anatomy professor related to us in a lecture on the integumentary system that scales, feathers, hair, and skin all appear in the fossil record fully formed and, in spite of over 150 years of looking, no transitional forms have been found bridging them.

At this, my hand shot up and I asked, "Isn't this a problem for evolution?"

He answered, "Not at all. Evolution is a fact; thus, the transitional forms must exist, and we just have to keep on looking to find them. It could be that the transitional stages were very unstable compared to the scales, feathers, skin, and hair, and that evolution from one to the other was very rapid and left very few transitional forms."

"But if the scales and feathers were so stable, why did they evolve into other forms of the integumentary system?" I asked.

Ignoring my question, he answered curtly with the argument, "You cannot use facts to challenge a theory that has been proven a hundred times over," and that was the end of that discussion!

At this, I did not say anything, but thought to myself that it seemed to me evolution is a science stopper. Thereafter, I concluded this to myself more than once.

Some Things I Pondered

After these class discussions, a major question I had was, "How can mutations cause evolution from molecules to man when we know that mutations are virtually always either nearly neutral, meaning slightly damaging, or deleterious? These near-neutral mutations add up, and eventually, when enough of them accumulate, they cause disease such as cancer, often leading to death, and eventually result in genetic meltdown. If the mutations are harmful to the degree that they cause early death, they would be less likely to be passed on to future offspring. Thus, in this case, they would be eliminated from the gene pool." I would think about this same question for the next two decades. This idea of mutations would be a topic I would spend several years working on in medical school, and for much of the rest of my career.

I learned in college that the two pillars of evolution are: *mutations* that produce variety, which is then pruned by *natural selection*. I would later conclude that natural selection is the *opposite* of evolution. It *reduces* variety and eliminates many life-forms. It does not produce new life-forms, but, in the end, only causes the extinction of many different varieties of life. The fossil record is very clear: thousands of animals have become extinct. Examples include woolly mammoths, saber-tooth tigers, the dodo, the passenger pigeon, the Moa of New Zealand, and the Great Auk. Mutation, likewise, is the opposite of evolution. It damages the genome, causing disease and death, not life or improvement.

Furthermore, no evidence exists of the evolution of a new life form above the genus level. Although rare, a few mutations are beneficial in extremely limited circumstances. Conversely, 99.99 percent are either nearly neutral or detrimental. The near-neutral mutations accumulate, eventually causing genetic meltdown and species extinction. It would take me almost twenty years to document the fact that the whole core basis of evolution is, without question, unequivocally wrong. It does not cause the ascent of life, but the descent of life, not evolution, but the opposite, devolution. But that's getting way ahead of my story.

I continued to read about evolution for the rest of my life. Much later, I read an article by Harvard Professor Richard Lewontin, which illustrated my concerns back then. He wrote: "Natural selection explains nothing because it explains everything." He added that the reason for this explanation is that a bird species

> with a small bill may evolve a larger bill size, because some aspect of the environment has changed so that large-billed birds now have more offspring. … Such a theory can never be falsified, for it asserts that the only politically correct answer to this question is that some environmental difference created the conditions for natural selection of a new character. Thus, the failure to find the environmental factor proves nothing, except that one has not looked hard enough. Can one really imagine observations about nature that would disprove natural selection as a cause of the difference in bill size? The theory of natural selection is then revealed as metaphysical rather than scientific. Natural selection explains nothing because it explains everything.[5]

Lewontin concluded: "Natural selection of the character states themselves is the essence of Darwinism. All else is molecular biology."

---

[5] Lewontin, R.C. 1972. Testing the Theory of Natural Selection. *Nature*, 236(5343):181–182.

Chapter 2: On the Path to Medicine

When I was fifteen years old, my grandmother died of breast cancer. Her enormous pain and suffering made a lasting impression on me, especially because I was very close to my grandmother. She was the rock in my life, possibly partly because she was indulgent to a fault. As I grew up, she made me feel that I could do nothing wrong. Her death set me on a course to find a way to help diagnose and treat cancer. From that point on, I wanted to go to medical school, but I also always wanted to be a musician. I was torn between these two options.

While studying violin at the world-famous Juilliard School of Music in New York, I triumphed over nearly 100,000 other applicants and was awarded a Ford Foundation Scholarship at age sixteen. What subject to study in college was a very difficult choice for me because I loved music and was already playing professionally. Or should I pursue medicine, which requires a lot of work and eight years of schooling in challenging academic subjects?

The beautiful picture of my grandmother that sat on my desk clinched my decision—it would be medicine. I accepted the scholarship that enabled me to attend Harvard to complete a degree in cell biology, the basis of all biology.

After I completed my undergraduate work in cell biology, I started working on my MD at Harvard Medical School. At times during my medical school study, my anatomy professor, an outspoken fan of Stephen Jay Gould, exclaimed in class about some human body structure or system he was lecturing about. "Whoever designed this structure (or system) sure knew what he was doing. It works marvelously." He never indicated who he thought the designer was, nor did anyone in the class ask. Evolution was rarely ever mentioned for my MD degree, even at Harvard. The cause and curing of disease, though, was our major focus and was often our only focus.

When I was still a senior in medical school, I began dating Pamela Beckels, a fellow medical student. She was a petite brunette, cute as a button, and had never seen a "B" grade in her entire undergraduate college career. Although only in her first year of medical school, we were close to the same age. We soon studied together almost every evening. As a typical young, healthy male, I was crazy about her and started to ignore my studies. All I now wanted to do was indulge in entertaining activities, like going to movies and concerts, with her.

Neither one of us wanted to break up, so we decided to marry while we were both still in medical school. We could then live together and save a lot of money. We were married by a justice of the peace and soon settled into married life. We were a team, studied together, and, as two *could* live cheaper than one, we had fewer money worries. My grades also soon shot up. When in medical school, Pam went into pediatrics, where she remained working full-time until we started our family.

I graduated at age twenty-three, one of the youngest MDs in recent Harvard history. I did my internship, then completed a residency and advanced training in immunology at Stanford University Medical Hospital. I selected immunology because I had concluded that the immune system was the key to both preventing and curing cancer. My wife also transferred to Stanford to finish her MD degree. After practicing as an immunologist for a year, I decided I wanted to go into research and went back to complete a PhD in cell biology at Stanford Medical School so that I could go into cancer research and fulfill my cancer research dream.

During my entire career for the two degrees I earned in medical school, I listened very carefully in all of my classes for any mention of Darwinism. When completing my PhD work at Stanford, evolution was also rarely mentioned in any of my classes, except to note that some genes were "evolutionarily conserved." The comment that some genes were "evolutionarily conserved" means only that the gene is very similar in both so-called primitive and advanced life-forms. This is logical because all molecular structures are irreducibly complex. Only certain designs work, and minor changes often cause organ or cell failure or disease.

The genes that produce protein units involved in, for example, the Krebs cycle, which converts the energy stored in our food into energy used to charge our cellular currency, must have a high level of similarity in all life-forms. Complex protein structures such as ATP synthase function in the cell to convert ADP to ATP by adding a phosphate radical to the ADP molecule. All lifeforms must have many other parts, not just ATP synthase, to function, including RNA, tRNA, and ribosomes, to name a few. Aside from this, in my over 135 semester hours of coursework—mostly in molecular biology and related areas of biology—the subject of evolution was never mentioned.

At Stanford, when I was working in my lab for my embryology class, I remembered my high school instructor teaching Haeckel's theory that ontogeny recapitulates phylogeny. So, I selected the specific embryos that were pictured in our high school biology textbook to examine under the microscope. I found that they all looked very different from those in our textbook. I was stunned! As an embryo, we do not go through the evolutionary stages as Haeckel's theory teaches. This would be the first of many examples where I discovered that some icon of evolution was wrong, or worse, deceptive.

I Began Working as a Professor

After I completed my PhD from Stanford, my first academic position as a professor was at Cornell University in New York. My wife practiced pediatrics in a large clinic in Ithaca, and, by then, we had our first child, a boy we named David. I remember looking at my newborn son in the hospital and thinking, "I am now a father, and this child has changed me forever." And he did.

Being a father also made me think more about the scientific problems of Darwinism. Two years later, we had our second child, a girl named Abby. I spent at least one full day a week with the kids, reading to them or playing games in the evenings with the family. I relished watching my two children grow up. Their little personalities were very obvious at a very young age. My wife's parents, now retired, lived close by, so they gladly took care of our babysitting needs. We had a lot of help from my in-laws, allowing both of us to have careers. It was a perfect fit. It also allowed me the required time to do my cancer research.

At first, my career went exceptionally well. I published many scientific articles in the most prestigious journals in science, including *Nature, Science, PNAS,* and *Cell.* I had just turned twenty-nine and was nominated by a colleague for a Nobel Prize and was told that if I kept up my productivity, I would be a shoo-in. I had obtained several million dollars of grant money to fund my research, which supported an average of six graduate students, and I was publishing two to three scientific papers a month.

I was a full professor in only four years, one of the youngest in the history of the university. After I was on the Cornell faculty for five years, I took an early sabbatical at the University of Michigan (U of M) in Ann Arbor, Michigan. I had a year to think, and do research, and teach a few classes at U of M as a guest lecturer.

I also got to know several professors in the science department who I learned also had serious reservations about Darwinism, which reinforced my doubts. After long, late-night conversations and a few drinks, their doubts about Darwin would often surface, and often only then. They accepted evolution but rejected the mutation and natural selection theory as the means causing evolution yet could not come up with a more plausible theory of biological origins. I did meet a professor at the University of Wisconsin Medical School, Dr. Bruce Lipton, who was open about his "heresy." He wrote a best-selling book on epigenetics, the influence of nongenetic factors on genes. In this book, he acknowledged that to

> the chagrin of my US faculty colleagues, I returned to Wisconsin [from my sabbatical out of the country] … bent on challenging the sacred foundational beliefs of biology. I even began to openly criticize Charles Darwin and the wisdom of his theory of evolution. In the eyes of most other biologists, my behavior was tantamount to a priest bursting into the Vatican and claiming the Pope was a fraud.[6]

While at the University of Michigan, I also enrolled in law school. Compared to medicine and cell biology, law school was easy for me. The University of Michigan law school taught law by the case study method. We read and analyzed a lot of cases and regurgitated them in class and on exams. The professor would pick

---

[6] Lipton, Bruce. 2005. *The Biology of Belief.* Santa Rosa, CA: Elite Books, p. 25.

one of us and ask, "How was *Williams v. ACLU* decided and why?" I always had a very difficult time accepting injustice, so I took a lot of classes related to civil rights.

Counting my sabbatical year and the summer before and after it, I was able to complete most of my coursework in a year and a half and took a semester off from Cornell without pay to complete my law degree. Not long after, I took the New York bar exam and passed with excellence. I am now a licensed lawyer! I returned to Cornell and was able to teach a few classes on genetics and the law. Life was good, very good, that is, until my Darwin heresy surfaced. Then chaos broke loose, and my life changed forever.

Chapter 3: I Become an Out-of-the-Closet Darwin Skeptic

As time went by, I became increasingly skeptical about Darwinism. It took me almost twenty-five years of reading and research, starting in third grade when I was introduced to evolution, before I was confident enough to openly come out of the closet. Ironically, one very early factor, aside from the questions I had in college, was a required reading book my daughter brought home from day school. The teacher's introduction noted that the text was written in an engaging style to maintain the children's interest so as to teach what the author described as "the all-important theory of evolution that was central to all science."

That claim surprised me, because most of the science books we used at Harvard, the University of Michigan, Cornell, and Stanford rarely, if ever, mentioned Darwinism. The kids' book consisted mostly of pictures, and the teacher's section explained that the text gave only a birds-eye view of the molecules-to-human evolution theory. The next grade's text on evolution, the text explained, would go into much more detail. The text stressed that the "subject was important because evolution tells us where we came from and where we are going"—the same questions many religions claim to answer. The book's text, which the teacher read to the class because it was written for third graders, was as follows:

Where did life come from?

Have you ever wondered where everything came from? Primitive man thought that some god created life. Now we know, thanks to one of the greatest scientists who ever lived, Sir Charles Darwin, that natural selection acting on mutations created everything except those things made by humans and animals. Natural selection is survival of the fittest. The stronger and more fit life-forms are more apt to survive than the weaker, less fit life-forms in the struggle for survival.

How did life start?

Thanks to energy from lightning, a group of molecules was able to move around faster. Soon, some molecules stuck together, forming compounds. As these molecules got bigger, some molecules were able to protect themselves and formed a coat of fat around them. This was the first semi-cell membrane. Those first cells that were more fit survived, and the less fit compounds, those without the membrane, dried up. Soon, super cells called bacteria evolved. They had a clear survival advantage over the cells with the thin fat membrane.

Multicellular Life Evolves

Next, some cells stuck together to help them stay warm and work together, forming multicellular life, such as protozoa, which also gave them a survival advantage. This happened when some bacteria got inside some of these cells and became part of multicellular life-forms, which made them more fit than cells without bacteria inside them. Soon these cells, called eukaryotes (you kary oats), thrived because they were more fit than the cells without bacteria inside of them, called prokaryotic bacteria. These eukaryotic cells eventually evolved into worms, which then evolved fins to enable them to swim and became fish. Then the more fit, superior fish evolved legs and became amphibians.

Where did the animals we see every day come from?

Some amphibians, those that had the right mutations that made them superior to others, eventually evolved into reptiles. Natural selection does not sleep, and the reptiles that were more fit eventually evolved into mammals. Some mammals were more fit than others in their environment and evolved into primates such as monkeys.

Where did humans come from?

These monkeys evolved into (have your teacher pronounce these names for you) australopithecines, Neanderthals, Cro-Magnon man, Homo erectus, and Homo sapiens. In the survival of the fittest struggle, Homo sapiens prevailed. The other human races were all killed off, and now we are the only higher apes left! All the others became extinct, part of an early genocide perpetuated by the superior race.

All of these claims forced me to think about the many serious problems with the "lots of time converts molecules to man" theory. I had to wonder if all the higher evolved apes went extinct due to being less fit to survive compared to humans. If so, why were so many lower-level animals doing so well? Examples include bacteria, cockroaches, house flies, rats, rabbits, deer, and squirrels, and so many higher, more evolved animals were doing so poorly, such as most primates except humans.

My entomologist colleague told me that flies and cockroaches were not simple, but more complex than the most sophisticated human inventions, such as a rocket ship. As I learned more about science, the simple-to-more evolved assumption of evolutionists rapidly began to break down.

I also knew no evidence existed to support the putative rule that the so-called "less fit" animals inevitably go extinct. Rather, University of Chicago Professor David Raup, in his 1991 book *Extinction: Bad Genes or Bad Luck?*, reviewed the research documenting, contrary to the evolutionary claim, that extinction is often far more a matter of bad luck than bad genes.[7] The reason for the demise of most animals can often only be speculated, but evidence exists that chance was a major reason. Although the balance in nature shifts periodically to accommodate changes in the environment, balance has existed for as long as life has been on Earth. Nor do the reasons behind most known modern extinctions relate to fitness, but rather, they are caused by humans.

I soon learned that my conclusions about evolution were not politically correct, and, in time, I realized that it would cost me a lot to come out of the closet. Yet, I could not deny the scientific evidence. The research was perfectly clear, showing that evolution, defined as from molecules to humans purely as a result of time, chance, mutations, genetic damage, and natural selection, was scientifically impossible. In other words, the evolution theory that teaches that humans got here "from goo to you by way of the zoo" is false.

I soon began submitting articles to scientific journals that carefully documented the problems with this theory. Most were rejected. One editor said that my article was an excellent, well-documented, and well-reasoned review of a major problem with evolution, but I had only documented a problem; now I had to attempt to solve it. When I solved the problem, he said he would be honored to publish my research study.

He added that his journal does not publish articles documenting problems with evolution, only solutions. In other words, molecules-to-man evolution is true, and any objections to the theory cannot falsify it, but are only problems yet to be solved because the theory is fact (which I interpreted as a theory held dogmatically) and not to be questioned. For example, when I pointed out that the fossil record shows no evidence of fish evolution from non-fish, their response was "we just need to keep looking. We will eventually find some of the transitions because evolution is true, so they must exist somewhere."

Thus, the links *must* have existed; we just have not found them yet. But we will find them someday, or we will find out why we have not found them yet. The transitional forms are there, or were there, or we must come up with a reason as to why they are not found, such as punctuated equilibrium theory, a theory used to explain the lack of evidence due to preservation problems. This theory concludes that most animals stay the same for eons, but when changing from one animal type to another animal type, they are unstable and evolve very rapidly, thus leaving very few fossils. The theory speculates that very few fossils would be formed during this time because evolution would have proceeded far too rapidly to have left many behind.

I knew both of these reasons were unsupportable but the evolutionists' reason that because evolution was true, the transitional fossil *must* have existed, or there must be very good reasons for their absence in the fossil record. The alternate

---

[7] Raup, David. 1991. *Extinction: Bad Genes or Bad Luck?* New York, NY: Norton.

conclusion that transitional forms never existed because evolution never happened was unacceptable. This is not science, but dogmatism. I felt we should go where the evidence leads. In contrast, they saw the whole world through their evolutionary glasses that distorted everything.

Chapter 4: A School Lecture

After thinking about it, I decided to inform my daughter's teachers about my concerns. She came home with the illustration her teacher gave involving a set of pictures of the evolution of Mickey Mouse from a skinny cartoon in 1928 to the picture seen today of a very different Mickey. This, she showed, was just how humans evolved from a bunch of molecules to humans.

"It seemed convincing to me that how Mickey evolved is just like how we do," she said with a serious look on her face.

"Not really," I responded.

"Why not?" she asked.

"Because the changes were the result of intelligent design due to marketing research," I answered.

When I approached the teacher, who had a degree in elementary education with only one class in science called science methods, she said she would have to ask the principal. As I expected, the lawyers got involved. They decided that I could give a presentation to the entire school body, "but nothing about creation or against evolution, otherwise, you understand, we risk a lawsuit." So, I decided to present the raw theory of evolution in simple words. Another Cornell professor was also invited, so that he could be with his child, who was also a student at the school.

In preparation, I did some research on the 1987 Supreme Court case on creationism, *Edwards v. Aguillard,* that dealt with the constitutionality of teaching creationism in public school classrooms. On page 587, it, in fact, stated *exactly the opposite of what the school lawyers claimed,* namely, the court ruled teachers have the *flexibility "to supplant the present science curriculum* with the presentation of theories, besides evolution, about the origin of life."[8] So I thought I would do just that.

After the school's student body was assembled and I was introduced, I began by asking the 400-plus students assembled, "What is Darwin's theory?" "To make it simple," I explained that evolution teaches: "humans are the result of millions of mistakes called mutations, or damage to the genome. The genome consists of all of your genes, and your genes are the blueprint, or plans, that construct your body." I then added, "We humans first started as simple molecules, such as hydrogen," which I demonstrated by a balloon filled full of hydrogen that I caused to explode with a match. That sure got the kids' attention. "Then some molecules reacted like the hydrogen did when I set it on fire, and we got more complex molecules, such as carbon dioxide (that is what you exhale) and methane (common natural gas that we use to heat our homes). Next, molecules got together to form cells, which evolved, first into simple organisms, of which we have no idea what they were, because even

---

[8] *Edwards v. Aguillard,* 482 U.S. 578.

the simplest bacteria are more complex than the most complex machines made by humans. Then the amoeba evolved, and, due to even more damage to its genes (which somehow was selected by the survival of the fittest law), evolved into fish, then amphibians, reptiles, mammals, primates, and, last of all, humans evolved."

As an afterthought I added, "Natural selection selects, meaning the stronger, more fit life-forms adapt to their environment, and those that do not die out—It's eat, or be eaten. In other words, life evolved from hydrogen to people, or 'from the goo to you by way of the zoo' (that got a hardy laugh out of the students), all by the accumulation of mistakes or damage in their genes and the weak dying off. Thus, we are the result of billions of mistakes. And there you have it!"

After a pause, I added, "Do you know someone who is sickly? These people are often selected out, meaning they die, and so cannot pass on their genes to the next generation. This is evolution."

A bright boy sitting in the front row asked, "How do we know all of this?"

I answered his question, to make it as simple as possible, as follows: "We just line up a bunch of life-forms that exist today from the simplest to the most complex, and since scientists can only give naturalistic explanations, they reason it must have happened like the line-up we did, and that's essentially it! Oh, we look for fossils that fit our theory, and every now and then we find a few new candidates. If their dates in history seem to fit the evolution scenario, we say we have found a missing link, proving our theory."

The boy humbly, but sincerely, asked, "What about God? My father is a doctor and taught me that God designed life."

I was beginning to think that he was a plant! I answered, "The lower courts have consistently ruled that it is unconstitutional to bring up that explanation. This means it is against the law and not allowed in school."

He answered, "That's dumb! What if intelligent design is true?!"

I responded, "We still can't talk about it in a public school. It is verboten! Only naturalistic explanations can be covered in state schools. It's called separation of church and state. The state cannot establish a church. It's part of the First Amendment."

This same student then asked: "But talking about the belief that God created life is not a church." He asked with a frustrated look on his face, "What does describing a worldview have to do with establishing a church?" Later, the teacher informed me that he was a straight-A student and his father was a famous surgeon.

I told him, "I am sorry, but the school's lawyers told me in no uncertain terms that I cannot broach this topic."

He looked dumbfounded, and so did many of the other students.

After the presentation, my Cornell colleague asked me, "Don't you think you oversimplified the theory somewhat?"

I answered frankly, "You have to simplify it so that elementary students can understand the big picture."

"But how you explained it sounds like a fairy tale, and not even a very believable fairy tale at that."

"True," I noted, "but is what I said wrong?"

"No, not really. You just explained it so that it sounds unbelievable."

"Then how should I explain it?" I asked.

"Just explain that humans and chimps evolved from a common ancestor, like the human-looking fossil Lucy, and the same is true of all life. When explained that way, it is more believable."

"But do you want the whole picture, or just isolated parts, to make it more believable?"

"The key is to present it in such a way that it sounds plausible. You want to convince the kids that evolution is how we got here, and that no God created us. Remember, some of the children are creationists, like the kid who asked all of the questions, and you want to get that idea out of their heads as soon as possible. Start young, teach them well, and they will be Darwinists for life!"

"But is your approach honest?" I asked.

"In the cultural war, the end justifies the means," he answered. "We have to get the "god" explanation out of their heads and the science explanation to replace God in their heads as soon as possible. The younger the better."

"But why?"

"As Woodrow Wilson once said in an address to the New York City High School Teachers Association, 'The purpose of a university should be to make a son as unlike his father as possible.'[9] Well, the same is true of all schooling, even elementary level schooling."

I then asked what I thought was a logical question: "Why not let the students have their own beliefs? Why push Darwinism down their throats?" I just got a funding support letter that reveals their true motivation. Let me read it to you:

> Despite the scientific advances of the past few centuries, the world remains haunted by irrationality and superstition. Belief in the supernatural abounds, as evidenced by popular entertainment and the growth of megachurches. But to cope successfully with the existential challenges that civilization faces, including pandemics, global warming, and rampant overpopulation, we must educate a generation that is grounded in reality, not delusion.
>
> Every developing mind wrestles with the mystery of how we become who we are. *Before Darwin, the only sensible answer was that we were designed by a supernatural creator.*

---

[9] In "The University's Part in Political Life," 13 March 1909, in *The Papers of Woodrow Wilson* 19:99.

Now we know better; we evolved from more primitive forms. But this complicated idea is not self-evident. It is hard to understand and must be taught by teachers who have the knowledge, skills, and teaching aids to do so.

This is an essential task, because a scientific understanding of our origins is fundamental to a realistic worldview. And that is the only way we will be able to cope with the existential threats to civilization.

Fortunately, the Teachers Institute for Evolutionary Science (TIES) is up to the task, and I am committed to giving it support to reach as many teachers as possible. I hope others will join me.

I will match all donations made to TIES—up to $20,000.

Thank you in advance for your support of today's students (who will be tomorrow's scientists)!

Best wishes,

Lawrence I. Bonchek, MD, FACS, FACC, retired Cardiothoracic Surgeon

He ignored the fundraising letter, so I added I thought teaching the whole story was logical, to which he answered firmly, "Because evolution is central to all of the sciences. Everything evolves. It is the scientific explanation of where we came from, why we are here, and where we are going, questions that religion tries to answer, but the answers science gives are factual and true; those answers that religion gives are neither factual nor true."

"So where are we going with this?" I asked.

He responded, "Obviously, we all will die, and that will be the end of our existence. In time, all life will die and the whole universe will also die in a Big Crunch, the opposite of the Big Bang, and that will be the end of all life, everywhere. Oh, sorry, the latest theory is that entropy will destroy the universe and all that exists will forever expand, forming a cloud of thin dust that is close to a vacuum, which will pervade the entire universe."

"Should we tell that to students?"

"Of course, we should. That way they will live for the moment and enjoy each day as if it were their last. I see no gain in teaching them fairy tales like Christianity. This life is all you have, so you better enjoy it, and make the best of the time you have on Earth."

"And we should tell this to second graders, the grade your son is in?"

"Heavens, no. They're too young. But we should help them to accept evolution, and later they will realize the implications of the theory. We have to introduce it when they are very young, then build on it as they progress through school. I would say only in high school should we teach them all of the clear implications of evolution that I just noted."

"So, how should we teach it when they are young?" I asked.

"Let me give you another example to make it simple and believable," he added, "the evolution of dogs. Show a chart of dog evolution from the wolf to the 339 breeds of modern dogs existing today. I find that a very effective way to teach evolution. Then tell them, "See, evolution is a fact, and they will then easily generalize this example to all life. Wahoo, you got 'em! From a single cell to humans, all by mutations and natural selection!"

"But they are far more interested in humans, not dogs."

"First, convince them that evolution is true, then they will more easily accept human evolution. For this reason, I would not talk about human evolution until much later in their schooling.

It was then that I fully realized what this war was all about, namely indoctrination, in essence. "Evolution" of dogs is actually an example of the loss of information, not evolution due to the gain of information. In each new dog breed, there is a loss of information that had originally existed in the wolf kind. This example actually hurts their theory and does not help it. It reminded me of the Darwinist biology professor Bora Zivkovic at Wesleyan College, a Methodist College ironically, who endorsed teaching students "inaccuracies" that are "wrong" if doing so enabled educators to "help them accept evolution." When I got home, I checked the reference. His exact words were:

> You cannot bludgeon kids … (or insult their religion, i.e., their parents and friends) and hope they will smile and believe you. Yes, NOMA [the idea that religion and science are both valid but are in different spheres of knowledge] is wrong, but it is a good first tool for gaining trust. You have to bring them over to your side, gain their trust … And on that slow journey, which will be painful for many of them, it is OK to use some inaccuracies temporarily if they help you reach the students. If a student … goes on to study biology, then he or she will unlearn the inaccuracies in time. If most of the students do not, but those cutesy examples help them accept evolution, then it is OK if they keep some of those little inaccuracies for the rest of their lives.[10]

"Little inaccuracies," I said to myself. These are lies just like the little white lies that destroy a marriage or a political career.

I then recalled that Charles Darwin wisely counseled, "A fair result can be obtained only by fully testing and balancing the facts and arguments on both sides

---

[10] Why teaching evolution is dangerous. ScienceBlogs, http://scienceblogs.com /clock/2008/08/25/why-teaching-evolution-is-dang/, 25 August 2008.

of each question."[11] The next day, when I repeated Darwin's words to Professor Thomas Reed, he shot back, "Was that some lame-brained creationist that said that? You know, saying that in a classroom is unconstitutional," he responded, defensively.

"It was Charles Darwin who said it," I answered, hoping to pique his interest.

"Yeah, right. Sure it was. You lie!"

"It is in the 1859 edition of *The Origin of Species*. I just checked the reference last night."

"I doubt that very much" he said emphatically!

"Why don't you check the reference?"

"I will not bother because Darwin would never have said something so stupid about evolution. There is no other side and there is no controversy about evolution. It is fact!" he shot back with firm resignation to end the discussion.

I asked him later and found out that he never went to the trouble to check the reference, adding "I couldn't care less who said it. It is wrong."

---

[11] Darwin, Charles. 1859. *The Origin of Species*. London, UK: John Murray.

Chapter 5: Thinking About the Fossil Record

After much reading, I learned that, surprisingly, the fact that most species appear fully formed in the fossil record was widely acknowledged by leading paleontologists. For example, Stephen J. Gould, professor of zoology and geology at Harvard University, in his *Natural History* article "Evolution's Erratic Pace" wrote, "The extreme rarity of transitional forms in the fossil record persists as the trade secret of paleontology.[12] The evolutionary trees that adorn our textbooks have data only at the tips and nodes of their branches." He added:

> The history of most fossil species includes two features particularly inconsistent with gradualism: 1. Stasis. Most species exhibit no directional change during their tenure on Earth. They appear in the fossil record looking much the same as when they disappear; morphological change is usually limited and directionless. 2. Sudden appearance. In any local area, a species does not arise gradually by the steady transformation of its ancestors; it appears all at once and "fully formed."

Furthermore, Niles Eldredge, chairman and curator of invertebrates at the American Museum of Natural History, noted that novelty in life "usually shows up with a bang," and this is why paleontologists have

> shied away from evolution for so long. It seems like it never happened. [We see only] occasional slight accumulation of change—over millions of years, at a rate too slow to really account for all the prodigious change that [must have] ... occurred in evolutionary history.

He added that when we

> see the introduction of evolutionary novelty, it usually shows up with a bang, and often with no firm evidence that the organisms evolved! Evolution cannot forever be going on someplace else. Yet that's how the fossil record has struck many a forlorn paleontologist looking to learn something about evolution.[13]

One day, when I read these quotes to a colleague in my department, Professor Herzig, he shot back, "Yes, but I know Eldridge, and he is an evolutionist."

---

[12] Gould, Stephen J. 1977. *Natural History*. "Evolution's Erratic Pace," p. 14, May.

[13] Eldredge, Niles. 1995. *Reinventing Darwin: The Great Evolutionary Debate*. New York, NY: John Wiley, p. 95.

I replied, "I agree, and thus the quotes are significant because they come from an evolutionist."

"I said he is an evolutionist!"

I said, "I know he is, and in the end, I strongly believe it is the evolutionists themselves that will falsify their own theory, and they have already largely done so now. They just cannot see the whole picture because their evolution glasses blind them to it."

"The whole picture? Why do you say that?"

"Because they have their evolution glasses on, which distort much of what they can see, and therefore they cannot see how all of the parts they have fit together falsify their worldview."

"Well, my comeback is that you have your creation glasses on and cannot see that the whole picture proves evolution."

"Not in my case. I come from an agnostic background, and let the evidence lead me, and my conclusions result from the evidence I see without any glasses on."

"I am lost."

I said, "Let me give you an example. We know that the cell is by far the most complex machine in the universe, and new research is showing daily that its complexity is even greater than we knew of the day before; thus, it is unlikely to have evolved. Despite this, my colleagues responded that this is just another example of the creationists' argument from personal incredulity. Richard Dawkins made this expression famous, and I hear this claim a lot."

"My response," Dr. Herzig shot back at me, almost yelling, "is to dig your heels in and get busy doing research. This is how you find the answer to how abiogenesis occurred, and how the cell first evolved."

Later, when I asked him about constructing a perpetual motion machine, Professor Herzig answered, "It's impossible to build one. In fact, the US patent office will not issue a patent for one even though they have no shortage of submissions. People have tried every possible design, and they all have failed. The reason is that you will always lose energy due to friction and air resistance."

I responded, "How do you know perpetual motion machines are impossible? Isn't this just an example of personal incredulity? You need to dig your heels in and work to figure out how to make one of these machines! You are just giving up."

"Look, we know that the laws of physics prevent such a machine from being built. You always have some energy loss due mostly to friction, air resistance, and such. You can get the friction level down close to zero, but never zero. Even in a perfect vacuum, which itself is impossible, some gas molecules will exist to absorb energy. You will always have some friction. As long as two parts are in contact with each other, friction exists," he responded, with a "got you" smile on his face.

I then added, "And the same thing is true of the cell. We know that abiogenesis is impossible, not due to personal incredulity or ignorance, but due to our extensive knowledge of chemistry and biology, just like we know that a perpetual motion machine is impossible," I replied, admittedly somewhat smugly.

"I don't buy that. The only alternative is creationism, and I will never accept that view, no matter what the evidence is," he answered.

"I say we should follow the evidence wherever it leads. Why rule out *any* explanation until we know for certain the answer?" I responded humbly. He did not appreciate my response, so I just dropped the subject.

As he walked away, he said, "Nothing in biology makes sense apart from evolution. Nothing! Got that? Absolutely nothing!"

The claim, "nothing in biology makes sense apart from evolution," now sounded like a broken record. This claim is false. Professor Steve Fuller surveyed two leading biological databases. In the 1,273,417 articles published from 1960 to 2006, he found a mere 12 percent used the term evolution, or its variant forms, in the abstract or keywords, and only 0.4 percent included the major concept of evolution, namely natural selection.[14]

---

[14] Fuller, Steve. 2007. *Science vs. Religion.* Cambridge, UK: Polity.

Chapter 6: Evolution True but Going Backward

A few days later, my peers and I again got into discussions about evolution. Professor Herzig was arguing that evolution perpetually propelled life upward and onward.

I responded: "In fact, it is clear, that, as mutations have accumulated, we are devolving, not evolving. And fully 99.9 percent of all mutations are near-neutral or clearly harmful. The accumulation of mutations leads to degeneration of the genome and mutational meltdown. Evolution is true, but going backward, leading to genetic entropy."

"Why don't you believe that natural selection is the answer to the cause of evolution"? asked Herzig."

I responded: "The issue I have with Neo-Darwinism is that it teaches that the source of novelty is the accumulation of random mutations in DNA in a direction set by natural selection. If you want bigger eggs, you keep selecting the hens that are laying the biggest eggs, and you get bigger and bigger eggs. But you also get hens with defective feathers and wobbly legs. Natural selection eliminates the weak and maybe helps, but it doesn't create; at best, it reduces the level of devolution." I added: "This reminds me of what Cornell's William Provine admitted: 'Evolution is not my friend. … Evolution cares nothing about me.' Meaning in my life comes from the people who care about me."

"We always regard Provine as somewhat eccentric," he answered.

"Maybe, but is evolution your friend?" I asked.

"I guess not, but it did create me!"

I said, "Let me read what the eminent scientist Lynn Margulis wrote in answer to the question. Professor Lynn Margulis was once asked, 'In view of the major objections to evolution, how, then, do you think the Neo-Darwinist perspective became so entrenched in science?' to which she answered:

> In the first half of the 20th century, Neo-Darwinism became the name for the people who reconciled the type of gradual evolutionary change described by Charles Darwin … in which fixed traits are passed from one generation to the next. The problem was that the laws of genetics showed no change, stasis, not change.[15]

"She added that the founder of the modern field of genetics, Gregor Mendel,

---

[15] Teresi, Dick. 2011. "Discover Interview: Lynn Margulis Says She's Not Controversial, She's Right." *Discover*, June 16. https://www.discovermagazine.com/discover-interview-lynn-margulis-says-shes-not-controversial-shes-right-1033.

showed that the grandparent flowers and the offspring flowers could be identical to each other. There was no change through time. There's no doubt that Mendel was correct. But Darwinism says that there has been change through time, since all life comes from a common ancestor—something that appeared to be supported when, early in the 20th century, scientists discovered that X-rays, and specific chemicals, caused mutations.

"Furthermore, Margulis wrote:

> Mendel found seven traits that followed his laws exactly. But Neo-Darwinists say that new species emerge when mutations occur and modify an organism. I was taught over and over again that the accumulation of random mutations led to evolutionary change—led to new species. I believed it until I looked for evidence.

"She then explained that the evidence is simply not there. Furthermore, she was asked about the famous beaks of the finch evolutionary studies of the 1970s: 'Didn't they vindicate Darwin?' The answer is no.

"Peter and Rosemary Grant went to the Galapagos Islands where Darwin did his original research to see speciation happening. They measured the eggs, beaks, etc. on finches year after year. They found that during floods or other times when there are no big seeds, the birds with big beaks can't eat. The birds die of starvation and go extinct on that island."

Herzig then asked, "Did the Grants document the emergence of new species?"

I answered, "The Grants saw the large-beaked birds going extinct, the small-beaked ones spreading all over the island, and being selected for the kinds of seeds they eat. They saw lots of variation within a species—but not evolution of new species. Some of these criticisms of natural selection sound a lot like those of Michael Behe, one of the most famous proponents of "intelligent design," and Margulis had debated Behe. What is the difference between your view and that of Behe?

Herzig answered, "The critics, including the creationist critics, are right about their criticism. It's just that they've got nothing to offer but intelligent design or 'God did it.' They have no scientific alternatives."

I then added, "The person questioning Margulis asked her one more question which was, 'Why do you have a reputation as a heretic?' Margulis replied,

> Anyone who is openly critical of the foundations of his science is *persona non-grata*. I am critical of evolutionary biology that is based on population genetics. I call it zoocentrism. Zoologists are taught that life starts with animals, and they block out four-fifths of the information in biology [by ignoring the other four major groups of life, such as prokaryotes] and all of the information in geology."

About that time, my dean, with whom up to this point I had always gotten along with very well, walked up to us and called me into his office asking, "Are you an evolutionist?"

"Well, that depends on how you define…"

"Look. It's a simple question, are you or are you not an evolutionist? Or do you deny the foundation of all science?"

"Again, it depends on…"

"No, it doesn't depend on anything. All I need is a simple yes or no. I am a simple man and want a simple answer. I will ask you again, are you an evolutionist?"

"You need to define…"

"I don't need to define anything. Got it?! For the last time, a yes or no is what I demand."

I could see that he was not reasonable about this simple topic, so with difficulty I managed to change the topic. I have had to do this a lot lately.

Chapter 7: The Moral Implications of Evolution

As I studied Darwinism, I became more concerned with the moral problems of this belief. I read the PhD thesis and other writings by Professor Provine's PhD student, Greg Graffin. Professor Graffin deals with the implications and moral problems of Darwinism by avoiding or rationalizing them, such as observing that evolution

> does not have a direction. It is anarchic, yet out of this anarchy have come biological entities of great sophistication and beauty. Many of our most important human features are not adaptations to prehistoric environments, and humans are far from evolution's crowning achievement.[16]

Graffin does not seem to realize that "anarchy" (random chance plus chaos) cannot produce "great sophistication and beauty" in life. Rather, only intelligence and design can do so. He concluded that "there is no ultimate reason for our existence" and believes that "the universe is made up of only four things: space, time, matter, and energy—and that's it." Graffin recognized that the moral implications of this are profound:

> By abandoning the idea that an intelligent designer created us, we can wake with each dawn and say, "What's done is done. Now, how can I make the best of the here and now?" Despite catastrophic tragedies, life has persisted in evolving new varieties of unimaginable forms. I find comfort in the narrative of evolutionary history ... I don't know if that feeling is enough to replace the solace of religion in the lives of most people, but it is for me.[17]

When working on his PhD, Graffin also interviewed, in depth, twelve prominent evolutionary biologists. One, George Williams, concluded, "If it's natural behavior, it's bad, it's evil," but Graffin added that he

> can't go so far as to characterize evolution as evil. Evolution is simply the way the biological world works, ... And once the reality of evolution is accepted, it has a strange and forbidding beauty. It occurs over time periods beyond human comprehension. It has created organisms of fantastic complexity.[18]

---

[16] Graffin, Greg, and Steve Olson. 2010. Anarchy Evolution: Faith, Science & Bad Religion in a World Without God. New York, NY: Harper Perennial.

[17] Graffin, Greg, and Steve Olson. 2010.

[18] Graffin, Greg, and Steve Olson. 2010.

Professor Graffin has clear objections to Orthodox Darwinism, especially natural selection, noting that evolution does not tend toward perfection. It depends as much—if not more—on cooperation and random chance as on competition. He devoted a whole chapter, chapter 3, titled "The False Idol of Natural Selection" in his book, addressing this problem. Graffin concluded that, as he studied nature, natural selection is either wrong, or at least far from the whole story, noting that evolution usually includes a description of natural selection and its role in the diversification of species. These accounts provide a mechanism for evolution and assume that the proffered explanation covers it all. He added that over the years he has

> become increasingly dissatisfied with natural selection as an explanation for all evolutionary change. And since I believe that dogma must be challenged wherever it is found—whether in religion, science, or music—I have spent time exploring the ideas of the iconoclasts who have examined natural selection critically. The result is a picture of evolution quite different from the standard textbook account.[19]

One of his concerns is that many existing life traits have little or nothing to do with survival. For example, anthropologists have for decades attributed certain physical characteristics to natural selection, which research has now documented as due to other factors. An example Graffin gives is that evolutionists

> have speculated that the epicanthic fold of the eyes in eastern Asians evolved to protect their Ice Age ancestors from the glare of the sun off blinding snowfields. Or they have said that the short, squat stature of pygmies was an adaptation to the equatorial heat of Africa (while ignoring the fact that tall, lean Africans lived just a few hundred miles away).[20]

Graffin then concluded, pensively, that

> almost all of these speculations have turned out to be unjustified and unverifiable just-so stories. Recent studies have shown that most of our physical features have changed randomly as modern humans spread from our eastern African homeland into the rest of the world.

He added that he has determined from his research that many human traits are "more a product of human sexual selection [due to culture] than of natural selection," adding that many human

---

[19] Graffin, Greg, and Steve Olson. 2010.

[20] Graffin, Greg, and Steve Olson. 2010.

traits are perpetuated simply because people find them attractive. The same process could apply to height, body size, body shape, and other human characteristics. All sorts of cultural factors affect perceptions of human traits, which means that culture can have a big influence on human evolution … the genes just follow along.[21]

He concluded that "the same process could occur in many other species. Birds, for instance, constantly engage in displays of their plumage and singing to attract the females." The fact is, if choice is involved, there exist many factors that could

play a role in causing a female to be attracted to certain behavioral and physical traits. Sometimes the seemingly fittest males just don't do it for a choosy female. And without successful reproduction, an organism's fitness is zero.[22]

He added that most mutations just shuffle the "deck of cards," and what macroevolution needs is lots of new and different cards, not just moving existing ones around the population. Even though Graffin is a committed evolutionist, I agree with many of his astute observations.

---

[21] Graffin, Greg, and Steve Olson. 2010.
[22] Graffin, Greg, and Steve Olson. 2010.

Chapter 8: Irreducible Complexity

I also did a lot of research on the interesting concept of irreducible complexity, the idea that all physical systems require a certain number of parts in order to function. One example often used is a mousetrap, which needs a base, a spring, a catch, and other parts. Without every one of these parts, it will not work.

Everything that needs two or more parts to function is irreducibly complex. Evolutionists, such as Ken Miller, claimed that the bacterial flagellum, the propeller-like tail structure that propels the bacteria through their watery world, evolved from the so-called "simpler" type III secretory system (called the T3SS system), which ejects liquids or other materials. But to form the basal body structure of the flagellum, a lot more parts are needed than exist in the T3SS system. Only about ten T3SS proteins are homologous with the flagellum; the rest are all new. Another problem with the argument is that the T3SS system is, at best, a sub-part of the flagellum. It forms only an anchor in the cell membrane and lacks the rotary engine or propulsion parts of the flagellum; it is a far *less* complex structure than the flagellar system.

The T3SS can't be the precursor to the flagellum because it is found in only a few bacterial groups and is thought by evolutionists to have evolved very *recently* compared to the flagellum. In contrast, the flagellum exists in a wide range of bacterial phyla and is believed to have arisen very *early* in evolution. So, the evolutionary data itself proves that the T3SS can't be a precursor to the flagellum.

The fact that one sub-part can perform some function doesn't refute the irreducible complexity concept; an evolutionary pathway is required to prove that it evolved. It is like saying that if my laptop's power cord can power my toaster, then, my toaster does not have an irreducibly complex core. The fact is, it still needs the power cord even if the cord can be used for some other electrical device. Thus, I concluded, this argument was a straw man test of irreducible complexity.

Chapter 9: The Fate of Darwin Skeptics

I soon became aware of many reports of the censorship of *Darwin Skeptics*. My brother, an undergrad at Harvard, told me that his biology professor informed the class that he, the professor, could not be involved in discussions about the problems of evolution, so he told the group to discuss it as a class, and he would sit by and not say a word. He never told the class why, and my brother thought it somewhat strange to tell the class this. They discussed evolution, and he told me it became very clear that some students had very strong reservations about Darwinism, for which they had valid reasons. More importantly, the Darwinists could not answer them, at least that's what my brother claimed.

Then my brother mentioned that he really never bought into the anti-evolution side, but after that class, he realized for the first time the stranglehold the science establishment and the university had on academic freedom in this area. He also concluded, after reviewing the textbook's arguments and those presented in class, that the case for Darwinism was, in his words, underwhelming.

In another case, when I spoke at a major American university, a postdoc cell biologist came up to me to relate what happened to her. She explained that she had raised doubts about Darwinism during informal conversations, and her boss asked her if she was a Christian and a creationist. When she said yes, he made it very clear that, if she ever spoke publicly about her belief, he would have to release her from all of her duties at the university.

Furthermore, he stressed, even if she completed her postdoc satisfactorily, she would not be able to publish, obtain grants, and likely would not be able to obtain a position in science. When she spoke to me about this experience, speaking rapidly with fear and tears in her eyes, she knew her career in science was likely ruined even before it began. She never had problems as an undergrad or a graduate student in cell biology, but did note that over and over, she would ask a question and would get an answer like, "it evolved that way," or something similar, which she regarded as a non-answer.

As time went on, I learned more about the fate of Darwin doubters at the hands of Darwinism's true believers. The following letter, which I received from a leading life scientist, even though it turned out well, is a typical example:

> The director of the scientific department for which I'm working is an Intelligent Design advocate and a devout Christian—so much for the claim that belief in God and Intelligent Design (ID) leads to the atrophy of rationality and analytical thinking skills.
>
> Because much of my work at the University involves highly classified military projects, I was subjected to an exhaustive background check

before being hired. I assume they Googled my name, in which case my involvement in ID and my conversion from atheism to Christianity would have surfaced.

I was hired because my boss, who made the decision, was also a Darwin Skeptic. If it was not for this fortunate event, I would no doubt be unemployed now.

Because he knew I was also an ID supporter, he came into my office one day. We then had a long discussion. When I told him I was once a Richard Dawkins-style atheist, he said to me, "Boy, were you misled!"

For obvious reasons, we have to keep all of our conversations private. I do not want my support for ID to jeopardize my career, and I'm not taking any chances. My new position is very challenging, interesting, rewarding, and pays well. And I want to keep it!

Another friend wrote about his work, and, along with other information, wrote:

A couple of months ago, I was hired by an aerospace R&D company because of my expertise in finite element analysis of computer simulations of inflatable structures. This is an extremely specialized discipline, and only a small handful of people in the world are knowledgeable about this field.

My current projects include simulations of supersonic and hypersonic (above Mach 5) inflatable aerodynamic decelerators. These devices are designed to decelerate space reentry vehicles from very high altitudes at extremely high speeds. I'm working with an extremely brilliant team of scientists but dare not reveal my views on evolution to them, even though my work has nothing to do with evolution.

Chapter 10: Testifying in Court

As the word got out about my heresy, I was asked to serve as part of the team of lawyers that was defending an academic freedom bill designed to protect the careers of teachers who critiqued Darwinism in class.

After the ACLU witnesses objecting to the bill were sworn in, as the attorney for the side supporting the bill, I asked questions of the ACLU attorney in court. The following is a paraphrased transcript of that questioning.

Radinow: Explain why you believe that government school teachers should not be protected for discussing the problems with Darwinism.

ACLU attorney: Good reasons exist to oppose this law, namely that teaching evidence against evolution is nothing more than backdoor teaching of creation. That violates the separation of church and state because creationism is a religion. Only two alternatives exist: either intelligence was involved in our origins, or it was not involved in the creation of the universe and life. One side is creation, the other evolution, and only the evolution side can be taught. All material on the other side must be censored, even that which simply does not support evolution, because this is teaching "backdoor" creationism.

Radinow: Could you define creationism?

ACLU attorney: I would say it is any material found in books, and journals, or magazines supporting creation or Intelligent Design.

Radinow: What about this secular college-level anatomy and physiology textbook? Should it be banned?

ACLU attorney: No, of course not. It is actually the textbook we use at the university where I teach.

Radinow: This secular textbook contains 678 pages detailing the intelligent design of the human body. To be consistent, human anatomy cannot be taught because it is the heart of Intelligent Design. How can we train nurses and doctors if this subject cannot be taught?

ACLU attorney: Well, anatomy and physiology can be taught, but a book teaching Intelligent Design, or a book advocated by creationists, cannot be used.

Radinow: Well, all anatomy books teach intelligent design from cover to cover. They all explain how intelligently designed the body is. Read one if you don't believe me. Are you arguing that the body is poorly designed?

ACLU attorney: Of course not.

Radinow: So, you agree ID is taught in A & P and should be taught?

ACLU attorney: Well, I will have to look further into this topic.

Radinow: Would all of the material published in *The Creation Research Society Quarterly,* or *CRSQ*, be illegal to teach in a public school?

ACLU attorney: Yes, of course it would. All of it.

Radinow: What about the *Journal of Creation*? Would it be illegal to suggest students in public schools should read articles from it?

ACLU attorney: Of course, it would. The courts have consistently made that very clear. Nothing in these journals could be taught in a public school. This material could be taught in Sunday school, but not in public schools, and, hopefully, it is not taught even in private schools.

Radinow: Can I read to you some of the titles of articles published in the *CRSQ* and *The Journal of Creation?*

ACLU attorney: That is fine with me.

Radinow: Do any Vestigial Structures Exist in Humans? The Rise and Fall of Haeckel's Biogenetic Law. The Life Clock and Paley's Watch: The Telomeres. Why Dawkins's Weasel Demonstrates Mutations Cannot Produce a New Functional Gene. Mad Cow Disease and the Enormous Complexity of Protein Folding. Does the Acquisition of Antibiotic and Pesticide Resistance Provide Evidence for Evolution? The Century-and-a-half Failure in the Quest for the Source of New Genetic Information. Does Gene Duplication Provide the Engine for Evolution? Could any of the material in these articles be taught in public schools?

ACLU attorney: If these articles are in creation magazines, I would not allow them to be used in public schools.

Radinow: What if the same material is printed in a non-creation magazine? I have a cell-biology book here, and it covers telomeres, mutations, protein folding, and antibiotic resistance. Could this book be used?

ACLU attorney: I would say this material should not be allowed to be taught in any government schools if it is in a creation book, but anything in a secular text can be taught.

Radinow: Thus, you would not allow anything related to genetics or cell biology to be taught?

ACLU attorney: I would not say that at all.

Radinow: That is what you just testified could not be taught in public schools.

ACLU attorney: I did not. I stated the material covered in the articles that you just read.

Radinow: But these are the areas covered in the articles I just cited!

ACLU attorney: If they are creationist articles, they are obviously not about genetics.

Radinow: Let me show you an article. (I showed him an article obviously on genetics.) Is this article on genetics?

ACLU attorney: (After looking through the article for several minutes) Yes, it looks like the article is all about genetics.

Radinow: So, it would be banned?

ACLU attorney: Only if it was written by a creationist or published in a creation magazine. If it were, the article must be biased or slanted.

Radinow: Do you think articles in evolution journals are unbiased?

ACLU attorney: Of course, they are objective.

Radinow: So, only articles in creation journals are biased?

ACLU attorney: Yes, I would say they all are. I do not trust creation material and refuse to read anything written by creationists.

Radinow: So, you censor not on content, but on the religion of the author?

ACLU attorney: If you want to put it that way, yes, I will censor based on the religion of the author. These authors have an agenda, and it may not be obvious, so the safe approach is to just not use anything written by a creationist or ID supporter.

Radinow: Is the content of these articles the typical material taught in Sunday schools and church sermons?

ACLU attorney: I don't know what is taught in church because I don't go to church. Never have and never will. I am an evolutionist and an atheist.

Radinow: Well, just take a guess, do you think these are Bible topics?

ACLU attorney: Probably not, but I would not know from personal experience.

Radinow: How about "The Darwinian Foundation of Communism?" Could the material in this article be taught in public schools?

ACLU attorney: No, that material cannot be taught in public schools.

Radinow: Do you believe students should not learn about the Darwinian foundation of communism?

ACLU attorney: No.

Radinow: Why not?

ACLU attorney: Because it will cause them to question the wisdom and truth of Darwinism, and we cannot allow that because Darwinism is the foundation of all biology. Nothing in biology makes sense aside from evolution.

Radinow: Could they read Richard Weikart's 1999 PhD dissertation, completed at the University of Iowa, titled *Socialist Darwinism* that documents how critically important Darwinism was in the communist movement?

ACLU attorney: No.

Radinow: Why?

ACLU attorney: I told you, because it could cause students to question the foundation of all biology, which is evolution and we cannot allow that.

Radinow: Even if the dissertation was accepted for a PhD degree at a major leading public state university?

ACLU attorney: No, I would not allow a teacher to teach from this source or to use this source in his classroom. We must not allow Darwinism to be questioned in the classroom in any way.

Radinow: What about ID supporter William Dembski, whose PhD thesis from the University of Chicago was published in a book by Cambridge University Press?

ACLU attorney: Likewise, I would not allow this thesis or the book to be used in a government school for the same reason.

Radinow: Could we teach that religion is the enemy of science and reason?

ACLU attorney: Of course we can, and I teach this idea myself because it is true.

Radinow: But that is teaching religion, is it not?

ACLU attorney: It is ok to teach *against* religion, just not *for* it, a decision that was clearly ruled by the court in Paloza and other court cases.

Radinow: And students should not question the concept of Darwin?

ACLU attorney: Heavens, no. The courts, such as the Dover, Pennsylvania, ID case, ruled that this is teaching backdoor creationism, which is unconstitutional. Also, there is no evidence against Darwinism, so the only reason to claim that there is, is an attempt to bolster creationism or ID.

Radinow: How about "The Galileo Myth and the Facts of History?"

ACLU attorney: Clearly, the contents of this article could not be taught in any public school.

Radinow: Why not, if the same general material is found in both books?

ACLU attorney: Because everyone knows that the Catholic church opposes science, and students should realize that religion is an enemy of both science and reason.

Radinow: Have you read any scholarly books or articles about Galileo?

ACLU attorney: No, I haven't.

Radinow: Then, how can you come to this conclusion?

ACLU attorney: It is common knowledge. All educated persons know this fact.

Radinow: Could common knowledge be wrong?

ACLU attorney: I suppose it could, but not in this case.

Radinow: Why not in this case?

ACLU attorney: Because the Christian church has opposed science from the beginning of time.

Radinow: Where did you learn that?

ACLU attorney: It was drummed into us in college. I read Professor White's two-volume book on religion's war against science, and also Draper's book on the same subject in college.[23]

Radinow: Let me give you two statements, and then could you please select which one is true. 1. The Church persecuted Galileo because of his theory that the Earth orbits the Sun. 2. The Church persecuted Galileo because he had annoyed some influential Jesuits! Which one is true?

ACLU attorney: I would say that the first statement is obviously true.

Radinow: Would it surprise you to learn that the author of these two statements was Professor Alexander Canduci, who labeled the first statement a "lie" and the second "truth"?

ACLU attorney: I am not a historian and have to rely on what my professors taught me in college and what I learn from the media.

---

[23] Draper, John William. History of the Conflict Between Religion and Science. Scotts Valley, CA: CreateSpace Independent Publishing Platform; White, Andrew Dickson. 1896. A History of the Warfare of Science with Theology in Christendom. Amherst, NY: Prometheus Books. Many editions are in print.

Radinow: Have you read any of these articles and books? (Reads the list of references from the *CRSQ* article.)

ACLU attorney: No.

Radinow: Would it surprise you to learn that all of these secular articles and books document the article "The Galileo Myth and the Facts of History" that you would not allow in public schools?"

ACLU attorney: I have not read any of these articles and books, so I would not know if they documented the claims made by this article.

Radinow: How would you know that church doctrine was the main problem Galileo had if you have not read scholarly articles about this case?

ACLU attorney: I told you; I just know from knowledge common to all educated Americans.

Radinow: Let me give you several more examples.

ACLU attorney: Fine.

Radinow: Ota Benga: The Pygmy put on Display in a Zoo. Could anything about Ota Benga be taught in public schools?

ACLU attorney: If it is in a creation friendly magazine, I would say that use of that material would be illegal. I told you the DeHart and John Freshwater court cases, and other cases as well, have decisively established the case law in this area.

Radinow: So, we should censor history; that is what you are saying?

ACLU attorney: No, of course not.

Radinow: Yet that is what you are doing! Is it not?

ACLU attorney: I am not! Creationism is not about history! It is all about religion.

Radinow: Why not? Is there some law that says creationists cannot write about history?

ACLU attorney: No, of course not.

Radinow: Then why can they not write about history?

ACLU attorney: Because they are not history scholars!

Radinow: If they have a PhD in history from a leading secular university, are they a historian?

ACLU attorney: I cannot imagine a creationist earning a PhD in any subject from a secular university, except maybe in religion.

Radinow: Why not? Thousands of creationists have PhDs, including me, from secular universities. What if I had an article in a non-creation magazine covering the same material as a creation magazine: Would it be illegal to use it in a government school classroom?

ACLU attorney: Yes, of course it would be.

Radinow: Why?

ACLU attorney: I told you, this would be teaching back door creationism, thus it would be unconstitutional to teach this material in all public schools."

Radinow: What about the article "The Exploitation of Non-Westerners for Evolutionary Evidence" or what about "Darwin's Ape-Men and the Exploitation of Deformed Humans?"

ACLU attorney: If it were in a creationist magazine, I would say no, it could not be taught in a public school.

Radinow: And if the same material was taught in a non-creationist magazine, what would you say?

ACLU attorney: If it were the same material, I would say that it could not be legally used or taught in any public schools.

Radinow: So, what you are saying is that anything about racism could not be taught in a public school.

ACLU attorney: What? Where do you get that conclusion? I did not say that at all.

Radinow: These articles both cover racism, and any article on Ota Benga would cover the same material as that in a creation magazine. A few books that cover the same material include … (He then cut me off.)

ACLU attorney: If these books implied that Darwinism influenced racism, they could not be taught or used in a government school as that would cause students to discount the unifying theme in biology.

Radinow: Why?

ACLU attorney: I told you; the reason is that it would cause students to doubt Darwinism and that would be supporting backdoor creationism and thus is not allowed in a public school.

Radinow: If an article printed in a non-creation science journal was reprinted in a creation journal, would that be illegal to allow students to read it?

ACLU attorney: Yes, of course it would.

Radinow: Why?

ACLU attorney: Because it would influence students to discount Darwinism, something that cannot be done in a public school.

Radinow: What if an article in a creation journal was reprinted in a non-creation science journal? Would that be illegal to use in a public school?

ACLU attorney: Yes, of course it would.

Radinow: So any article that *ever* appeared in a creation journal would be illegal to use in public schools?

ACLU attorney: I would say, based on past court case decisions, that it would clearly be illegal unless the teacher used it to discredit creation and I.D.

Radinow: What about the article printed in a non-creation journal titled "The History of the Human Female Inferiority Ideas in Evolutionary Biology.

ACLU attorney: I would think that would be OK to allow students to read.

Radinow: But if the article was written by a creationist, would it be OK to allow students to read it?

ACLU attorney: In that case, no, it would not be allowed.

Radinow: So, anything written by a creationist would be banned no matter where it was printed?

ACLU attorney: I would say, yes.

Radinow: Isn't this like cooties? Everyone touched by a person with cooties is tainted?

ACLU attorney: I would not say that.

Radinow: Do you know that Nazi Germany banned anything written by Jews, and you are doing the same thing—just substituting creationists for Jews.

ACLU attorney: I resent that implication. We in America today are not like the Nazi's.

Radinow: But isn't it true?

ACLU attorney: No, of course it is not true!

Radinow: But that is exactly what you are saying, is it not?

ACLU attorney: It is not!

Radinow: If the same material and conclusions that were made in a creationist's journal were published in a secular, non-creationist journal by an evolutionist, should it be banned?

ACLU attorney: Ok, I see your point. To be consistent, yes, it should be banned.

Radinow: In several cases, the court has ordered teachers to be monitored by recording and reviewing all of their lectures to ensure that they do not teach any of the material you say should be banned. Is this a good idea?

ACLU attorney: Yes, we must enforce the US Constitution and ensure that creationism, or anything against evolution, is not taught in public school classrooms.

Radinow: Who would you monitor, evolutionist teachers or creationist teachers?

ACLU attorney: Just creation-believing teachers.

Radinow: How are you going to know this?

ACLU attorney: I would have every teacher sign a statement swearing that they are either a creationist or an evolutionist.

Radinow: Isn't that illegal?

ACLU attorney: I would say, to ensure that information in favor of intelligent design or creationism is not taught, and only information in favor of Darwinism is taught, it would be a necessary exception so that we know which teachers to watch carefully to ensure that they do not present any evidence that causes students to question the proven fact of Darwinian evolution.

Radinow: But surveys show that about 90 percent of the American population is some kind of creationist.

ACLU attorney: Well, we will have to hire a lot of classroom monitors, won't we?

Radinow: Who will pay for this?

ACLU attorney: The taxpayers, of course.

Radinow: So, we will have to raise taxes, is that not correct?

ACLU attorney: Either that or cut back on unimportant subjects like physical education, music, and art. You don't seem to take the Establishment Clause of the Constitution very seriously, do you?"

Radinow: Well, when my folks went to school, they were taught creationism in science class, and most of them did very well. Religion has been taught in the public schools in America for over 300 years, actually until the late 1960s.

ACLU attorney: Obviously then, we did not follow the Constitution for over 300 years, did we?!

Radinow: So it took three hundred years to figure out that we did not follow the Constitution!?

ACLU attorney: I guess so.

Radinow: The article, "Controversy in Evolution" is in a non-creationist journal. Could that be taught?

ACLU attorney: No, there are no real controversies in evolution.

Radinow: None?

ACLU attorney: Yes, none. Evolution is a fact, so there are no real controversies in evolution. The debate is only over very minor points, like did humans first evolve in South Africa or Northern Africa, not over evolution.

Radinow: I have a whole book here published by a non-creationist publisher, written by evolutionists, titled *Controversy: Catastrophism and Evolution: The Ongoing Debate*, published by the leading science publisher Springer Verlag. Would you allow this book to be in a public-school library?

ACLU attorney: No, I would not!

Radinow: Why?

ACLU attorney: I just told you, because there is no controversy about evolution, and this book will no doubt adversely affect students' belief in evolution, even if it was written by an evolutionist. We cannot allow creation, or even any criticism of evolution, a foot in the classroom door.

Radinow: I have an ID book here on human junk DNA by a University of California, Berkeley, cell biology PhD,[24] arguing that good evidence exists for the view that most DNA is not junk (meaning useless DNA), and another article from an evolutionist journal documenting that almost all DNA is junk and has no function in the genome. Which one would you ban, and which one would you allow to be taught?

ACLU attorney: Of course, I would ban the ID book by the cell biology PhD and allow the article published in the evolutionist journals to be used in science classes because it is science.

Radinow: But if the 30 pages of references in the "DNA is not junk" book were, with very rare exceptions, all from evolution journals, would you still censor the ID book?

---

[24] Wells, Jonathan. 2011. *The Myth of Junk DNA*. Seattle, WA: Discovery Institute Press.

ACLU attorney: Certainly. The ID book is probably quote-mining. Look, there is no disagreement about evolution among scientists. It is a proven fact.

Radinow: What if the teacher selected several of the articles listed in the references section of the ID book to have students read. What then?

ACLU attorney: I would not allow them to read these articles. This was the ruling decided in the Roger DeHart court case, so my decision has a court precedent.

Radinow: In other words, you would censor everything supporting ID and everything not supporting evolution, but allow everything supporting evolution in the classroom?

ACLU attorney: If you put it that way, yes, that is exactly what I would do.

Radinow: Why? Are you in favor of censorship?

ACLU attorney: I would censor only everything that violated the Establishment Clause.

Radinow: How do articles on DNA violate the Constitution?

ACLU attorney: If they support evolution, they don't violate the constitution, but if they support creation, then they do, and are thus unconstitutional because this results in sneaking ID ideas into the classroom by the back door.

Radinow: So, any scientific fact that supports ID or creation is religion, and anything that opposes ID is not religion?

ACLU attorney: Correct.

Radinow: What about anything that supports atheism, would that be OK to teach?

ACLU attorney: Yes, as I said before, it would be allowed.

Radinow: Why?

ACLU attorney: Because it does not support theism.

Radinow: Isn't this blatant indoctrination?

ACLU attorney: Yes, but it is necessary to protect the constitutional rights of students.

Radinow: Where in the Constitution does it forbid teaching students facts about the scientific problems with molecules-to-man evolution?

ACLU attorney: The Establishment Clause. Teaching that there are problems with evolution is religion, thus unconstitutional. Besides, there are no real problems with Darwinism, only falsely claimed ones.

Radinow: You are saying that teaching students the problems with evolution is the state establishing a church?

ACLU attorney: Yes, the courts have repeatedly ruled such is the case.

Radinow: How do you justify it?

ACLU attorney: Easy. Evolution is a fact, and creation in all its forms is not only religion, but also dangerous. This is what the Council of Europe ruled in a 2007 report by Professor Lengagne. I have a copy of it here.

The ACLU attorney then took a copy off the desk and, after reading it, concluded that it says "if we are not careful, those who doubt Darwinism could become a threat to human rights" because their "modes of thought" attack the

> very core of the knowledge that we have patiently built upon nature, evolution, our origins, and our place in the universe. There is a real risk of a serious confusion being introduced into our children's minds between what has to do with convictions, beliefs … and what has to do with science … which may seem appealing and tolerant but is actually disastrous.

> Intelligent design, the Council added, 'does not deny a certain degree of evolution but claims' that the universe and all life is 'the work of a superior intelligence,' a view, they concluded, is a serious threat to human rights. The council concluded with the ominous warning that something must be done about those who are not Darwin true believers 'before it is too late.' They added that though

>> subtler in its presentation, the doctrine of intelligent design is no less dangerous [than creationism]. A detailed study of the growing influence of creationists shows that the discussions between creationism and evolutionism go well beyond intellectual disputes. If we are not careful, the values that are the very essence of the Council of Europe will be in danger of being directly threatened by the creationist fundamentalists. It is part of the role of the Council's parliamentarians to react before it is too late.

ACLU attorney: The highly esteemed scientists who wrote the Council report added that many scientists hoped that the "theory of natural selection would enable humankind to finally put an end to the theoretical foundations of 'religious obscurantism.'" The report then quoted a zoology professor who authoritatively proclaimed that science 'depends on methodological materialism' and the 'neo-creationist movement,' which mainly consists of the advocates of 'intelligent design' argues for the 'hypothesis' that a 'so-called superior intelligence' has intervened in human history.

This 'superior intelligence' is called God by most people, and the theory that He has intervened is called theistic evolution or BioLogos. The Council of Europe thus finds that only atheism is acceptable, and theism is outright 'dangerous.' The report continued, describing ID as nonscientific, and

> the supporters of intelligent design demand that their ideas be taught in biology classes alongside the theory of evolution. However, in 2005, the intelligent design creationists … suffered a setback when the Pennsylvania judge John Jones declared that the teaching of intelligent design in schools violated the constitutional separation of church and state. Similar criticism can be made about the "pseudo"-scientific character of the intelligent design ideas … the intelligent design ideas are anti-science: any activity involving blatant scientific fraud, intellectual deception, or communication that blurs the nature, objectives, and limits of science may be called anti-science.

I then resumed questioning the ACLU attorney.

Radinow: But how could the claim that ID is dangerous be true when 90 percent of Americans hold to some form of creation?

ACLU attorney: Well, they won't when evolution is intensively taught from K through 12 in every public-school class. We evolutionists are going to make sure of that!

Radinow: Do you know that in the Scopes trial, Scopes's attorney, Mr. Malone, on page 187 of the published trial transcript, said:

> The truth always wins, and we are not afraid of it… let the children have their minds kept open—close no doors to their knowledge; shut no door from them. Make the distinction between theology and science. Let them have both. Let them both be taught.

How could creation be dangerous when almost everyone believes it?

ACLU attorney: Well, many Americans believe a lot of things that are wrong, like astrology, the existence of UFOs, and ghosts.

Radinow: Then we should have anti-astrology lessons from K through 12?

ACLU attorney: We see no point in that.

Radinow: But you see a point in teaching anti-creation lessons from K through 12."

ACLU attorney: Yes, because creation beliefs are dangerous.

Radinow: And it is perfectly fine to believe in astrology or ghosts?

ACLU attorney: Yes, that material is okay in a public school, even though it has been proven false, because these ideas do not cause any harm to society. Only creation beliefs are dangerous.

Radinow: But why?

ACLU attorney: Because evolution is a fact, and the foundation of all science. Science has done more to help humanity than any idea or belief. Certainly, far more than religion, which has done very little good in society, but rather it has caused wars and holocausts for eons. That's why.

Radinow: But astrology is a disproven idea, so why is creation so dangerous and astrology not dangerous? And I want to add that it was not the ministers who developed the atomic bomb, poisonous gas, or the other horrible instruments of war that have killed so many innocent people. It was the scientists.

ACLU attorney: Because evolution is the science explaining everything, the whole universe and the Solar System, plus the origin of all life-forms, and astrology doesn't claim to explain everything, and also it is not religion as is creationism.

Radinow: How do you know evolution is a fact?

ACLU attorney: Because all credible scientists accept it. That's why.

Radinow: Is it not true that they all accept it, but most do largely because no other views are taught in the schools? They learn only one side, so logically they support the only side that they have learned in school.

ACLU attorney: No, they accept it because it is a fact. There is no reason to learn about the evidence against evolution because there exists no evidence against evolution.

Radinow: None? My research documents that there is an enormous amount of evidence against evolution, and, furthermore, science has shown evolution never happened and never could have happened.

ACLU attorney: Well, even if that is so, teaching this evidence, if it exists, which I doubt, is still teaching back-door creationism. We have only two choices. Either intelligence created the universe and all life in it, or the blind watchmaker did it. Only one view is legal in public schools, and it is not the intelligent design view. The only legal view is that the blind watchmaker did it all, meaning the blind watchmaker ultimately created everything!

After two weeks of testimony, and thirty-six witnesses, the trial ended. Six months later, we found that we had lost the case. The court ruled against protecting the teaching of any other view aside from the Darwinian worldview, because non-Darwinian views are unconstitutional—only naturalistic evolution can be taught in government schools, and all information against evolution must be censored. They thus ignored the 1987 Supreme Court case on creationism, *Edwards v. Aguillard,* dealing with the constitutionality of teaching creationism in public school classrooms, which on page 587 stated *exactly the opposite of what the school claimed.* The court ruled that teachers have the *flexibility "to supplant the present science curriculum* with the presentation of theories, besides evolution, about the origin of life."[25]

---

[25] *Edwards v. Aguillard,* 482 U.S. 578 (1987).

Chapter 11: The Opposition Intensifies

I soon began speaking at churches, Christian schools, and other community groups. The main problem I discovered was that many people had difficulty believing scientists actually accepted the "goo to you by way of the zoo" theory. Most importantly, they found it hard to believe that people were actually getting fired over rejecting, or even just questioning, this worldview. I learned to bring documentation, such as an article I authored on academic freedom. "If you don't buy my claims, read this article. It is well documented," I would often say.

Word of my lectures soon got back to the university. As a result, things were now very tense there and I was often challenged. One day Professor Goldstein, who was also Jewish and the son of a rabbi, told me that he became an atheist in college after studying evolution. Then he asked me,

> "Why don't creationists publish in scientific literature? Let me tell you why. Because you have no science behind your theory, it is pure religion based on faith only. If there were science behind your theory, you guys would publish lots of peer-reviewed articles proving that your theory is correct.

"The reason is," I replied, "once you are out of the closet as a Darwin Doubter, there is so much hostility that you will often be blackballed by the science establishment and find it difficult to publish anything in a science journal."

Goldstein replied, "That's a bunch of crap. No scientists would do that. Peer review is objective. If you have the evidence, you will be published. End of story. They reject your papers because your anti-Darwin junk is creationism and Intelligent Design, which is not science, but purely religion crap."

I had heard this argument over and over and frankly it made me mad, so I responded:

> The problem is, I am now no longer judged by my scholarly work, but by who I am. The science establishment ignores Martin Luther King's admonition that we should judge people by the content of their thoughts, not by the color of their skin or by a label such as "creationist." What evolutionists do is label someone and then reject their work no matter how good it is. I had no problem publishing until I was outed and now, even though I have grown enormously intellectually since I started publishing almost ten years ago, I find it harder to publish than ever before. This is nothing more than pure intolerance and bigotry.

"I don't feel one bit sorry for you. You are getting exactly what you deserve," Professor Goldstein replied nonchalantly.

"Let me give you an example," I added. "I have had scientific papers accepted, then when the editor found out I had serious reservations about the validity of Darwinism, the paper was never published and my inquiries about the paper went unanswered. I also have had a few papers friendly to Intelligent Design accepted and published, then every following paper I submitted to that journal after the ID article was published, was rejected."

Goldstein replied, "So? We all have had papers rejected. I have had several rejected, and I don't go off the deep end claiming discrimination like you do."

I responded, "Let me give you a few other examples. Every paper, six total, that I submitted to one journal, edited by an Intelligent Design friendly editor, was accepted, then the Intelligent Design friendly editor was fired as a result of the controversy that he created due to publishing an ID friendly book, and a few Intelligent Design friendly papers. Then a new editor was appointed. Since then, every paper I have submitted to this journal was rejected. I was told by the editor that another paper I published supporting irreducible complexity, resulted in over fifty subscription cancellations to the journal by college libraries. If a journal published an ID friendly paper, they could, as a result, find their journal blacklisted or worse."

"Like I say, you guys got what you deserve," he responded stoically. "Anyone who doubts Darwin should be blacklisted. They have no business teaching impressionable young people."

I responded, "Let me refer to a *Science* review article on a book that the prominent evolutionary biologist and science historian Stephen Jay Gould wrote, as reviewed by Lewis and his co-workers at Stanford University in *PLoS Biology*. As you know, *Science* is the leading science magazine in America. Gould argued that "unconscious manipulation of data may be a scientific norm" because "scientists are human beings rooted in cultural contexts, not automatons directed toward external truth," a view now popular in social studies of science.

In support of his argument Gould presented the case of Samuel George Morton, a nineteenth century physician and physical anthropologist famous for his measurements of human skulls. Morton was considered the objectivist of his era, but Gould reanalyzed Morton's data, and in his prize-winning book *The Mismeasure of Man*, argued that Morton had skewed his data to fit his preconceptions about human variation. Morton is now viewed as a canonical example of scientific misconduct. But did Morton really fudge his data?"[26]

---

[26] Lewis, Jason E., David DeGusta, Marc R. Meyer, Janet M. Monge, Alan E. Mann, Ralph L. Holloway. 2011. "The Mismeasure of Science: Stephen Jay Gould versus Samuel George Morton on Skulls and Bias." *PLoS Biology* (Public Library of Science Biology) 9(6): e1001071.

"Yes, scientific misconduct occurs, we all know," Dr. Goldstein acknowledged.

"It gets worse. The article's authors, Ralph Holloway and his students, concluded that Morton did *not* fudge his data, rather it was Gould who fudged his data when he was trying to prove Morton was guilty of fudging his data. Gould's own article proved his thesis that scientific manipulation is common in science!"

"Well, frankly I do not have much respect for Gould, so you're wasting your time with that argument," Professor Goldstein shot back. "All he was was a science popularizer. He did very little important scientific work in his career."

I then told him, "It seems to me that some professional jealousy is showing through here."

"Could be," he answered. "But evolution is a fact, and more evidence exists for it than for the round-Earth conclusion."

In response to my peers' claims that evolution is "fact, fact, fact," I opined that this is just not true. I found it interesting that a decade after Darwin sold out the first edition of his 1859 book *On the Origin of Species*, fully one-third of scientists did not accept it.[27] These scientists evaluated the theory and rejected it. How many accepted it later only because of social pressure?

I said, "Actually, one problem that I have had in researching the case against Neo-Darwinian evolution is that the evidence against it is so overwhelming that many creationists conclude that documenting it is a waste of time. I focused on this area because it's easy to research. Just identify a Darwinist claim, such as the vestigial organ claim, then research the literature. In this case, the claimed examples of vestigial organs, such as the thymus gland, the appendix, and the pineal gland, all have important functions, and are not useless or degenerate, and so you have falsified the evolutionists' claim.

"Some ID supporters feel that building a case against evolution is like publishing a journal titled *The Round Earth Journal* and spending one's life publishing articles in the journal proving that the Earth is round. It is without question round, and wasting time proving that it is round is not productive. Darwin Doubters feel that the same is true of evolution."

When I presented my findings to my colleagues, they claimed that an organ is vestigial if it merely has less function, and not that it is useless or close to useless. This is not what my textbooks claimed, and I realized that this was a new definition created to get around the fact that their vestigial organ theory had been disproven. Actually, no evidence exists that any claimed vestigial organ has *less* function than our ancestors had! In a few cases a *different* function has been claimed, but not less function. Thus, I concluded, researching evolution is not like researching the round-

---

[27] Hull, David. 1988. *Science as a Process*. Chicago, IL: The University of Chicago Press, p. 303; Hull, David, P. Tessner, and A. Diamond. 1978. Planck's Principle. *Science* 202:717-723.

Earth view because most scientists believe that evolution theory is valid, and none believe that the flat-Earth view is valid.

A few days later, when talking with a Catholic priest about the professors and scientists who lost their careers over doubting Darwinism, he said, "I do not see the problem with evolution. Pope Pius XII, the head of almost a billion Catholics, said in his 1950 encyclical that Catholic colleges must teach evolution 'in such a way that the reasons for both opinions, that is, those favorable and those unfavorable to evolution, are weighed and judged.' So, what is the problem with teaching both sides?"

"But this was a long time ago," I protested. "What do Catholics teach today?"

"Yes, it was a long time ago, but the Church's teaching has not changed. Pope Benedict XVI's address in April of 2005 said, 'We are not some casual and meaningless product of evolution.'"

"You cannot get any clearer than that!" the priest said with satisfaction written all over his face.

He then added, "And Cardinal Schönborn, archbishop of Vienna, in a *New York Times* article[28] said, 'Any system of thought that denies or seeks to explain away the overwhelming evidence for design in biology is ideology, not science.'"

"That is exactly what I have come to believe," I answered, noting how clearly he expressed my feelings.

"And that is what I believe, and most Catholics believe," the priest then added.

I added, "Still, the opposition to persons who are critics of Darwinism has been growing. At Iowa State University, over 120 faculty members signed a petition denouncing ID and calling on "all faculty members to ... reject efforts to portray Intelligent Design as science."

---

[28] Op-ed article titled *Finding Design in Nature* by Roman Catholic Cardinal Christopher Schonborn on the Catholic stance on Evolution, 7 July 2005.

Chapter 12: The Debate

I was set up to debate a leading scientist and evolutionist, Dr. Herman Rushfield, ironically, a former minister. The rules were simple: each of us had ten minutes maximum, then the other person would step in and speak for ten minutes. That rule did not last for even two minutes! Neither of us adhered to it. After the formal introductions, he went first.

Dr. Rushfield began with: "My argument is very simple: The fact is, every informed scientist accepts evolution. It is a proven fact, and no informed person denies it. Only ignorant fundamentalists who know nothing about biology, or science either, deny it."

I realized from his opening statement that, although the man is a leading biologist, he knew little about the whole evolution controversy. I felt that his ignorance would clearly hurt him in the debate.

"Furthermore, there is nothing to debate. Evolution is the core of biology," he added with a certain smugness. "I do not know of a single intelligent supporter of creation or intelligent design, not one, and I know thousands of scientists."

"First of all," I began, "I have a list here of over 3,000 mostly Ph.D. level scientists, all of whom are Darwin Skeptics."[29] I then handed him the list.

As soon as the list was in his hands, I continued, adding, "I personally know of many scientists who are not on this list because they know full well that coming out of the closet is a career ender, and some are so busy that they do not have the time to spend on the issue. Many of the scientists said they do support Intelligent Design, but are firmly in the closet, so you would have little inkling of their views on Darwinism even though you may work with these people."

"I doubt that," he said. "I would know if they were not evolutionists."

"Have you asked them?"

"No. Of course not. I just know they are evolutionists. Every scientist is."

"You mean you assume they are evolutionists?"

"No, I just know that they are."

I said, "Let me define evolution. Better yet, I will let Loren Eiseley who states that "the evolution of the entire universe—stars, elements, life, man—is a process of drawing something out of nothing. Out of the utter void of non-being, everything came to exist."[30]

---

29 www.rae.org/darwinskeptic.pdf.

30 Eiseley, Loren. *The Night Country*. New York, NY: Charles Scribner's Sons, p. 212.

I continued, "Another problem is that some creationists feel that the case against evolution is so overwhelming, they prefer to spend their time on developing the creation model."

"OK, what is some of your evidence?"

"As I read about fifty books a year, most of which are on science, it is clear that my conclusions are not all original. So, let me quote Professor Billington, a major well-known scientist, concerning natural selection, which is the heart of evolution. Professor Billington, in her leading textbook, *Understanding Ecology,* wrote that Darwin's struggle for existence and survival of the fittest is contradicted by the fact that a

> careful observation of a community shows that plants and animals live together in agreement. Every living thing has a will for life to go on. People have the mistaken idea that animals in natural communities are enemies. People believe that predators … are waiting—ready to snatch every passer-by. Ethology … the study of an animal's normal behavior, shows that this is not true. Both plants and animals appear to avoid direct competition with others if it might injure or kill them. [31]

"She added:

> Only when a community becomes overcrowded … does competition for food increase and bring danger. Because of the great "will to live," all living things seem to "cooperate" as well as "compete" with each other. When life is carried on normally in a community, the members live peacefully together. Often, one will alert another to a common danger. You can observe this in a city park, or street, or in a suburban garden where pigeons, sparrows, starlings, and other birds live together.

"Furthermore, she observed that there

> may be some 'pecking' at the smaller birds who come to feed, but it is not a fight to the death. Watch chipmunks or squirrels feeding and chasing, and you will soon realize that much of the chasing seems to be in fun. If a real fight should develop, the other animals in the area show great concern even though they are not involved.

"This is also what I have observed. As I read such statements, I sometimes noted to colleagues my observations that this was a problem for Darwinian gradualism. I also became increasingly confident about my conclusions as I read more and more about this subject."

---

[31] Billington, Elizabeth T. 1968. *Understanding Ecology.* New York, NY: Frederick Warne and Co., Inc., p. 52.

At this point, Dr. Rushfield just walked out! I was aghast and so was the audience, so I asked if we should all just go home. The audience responded, just finish your part. They wanted to hear what I had to say. So, I did.

I continued, "Ernst Mayr, Professor of Zoology at Harvard University, in his book published by Harvard University Press, wrote: 'Paleontologists had long been aware of a seeming contradiction between Darwin's postulate of gradualism … and the actual findings of paleontology.' He added that following phyletic lines through time revealed

> only minimal gradual changes, but no clear evidence for any change of a species into a different genus or for the gradual origin of an evolutionary novelty. Anything truly novel always seemed to appear quite abruptly in the fossil record.[32]

"Richard B. Goldschmidt, the late professor of genetics at the University of California, Berkeley, wrote in an article: 'The science which shows us evolution actually at work—paleontology—should be able to contribute important information'[33] to evolution, but it has not. In fact, the decisive steps in evolution are abrupt, without transitions, and when

> a new phylum, class, or order appears, there follows a quick, explosive (in terms of geological time) diversification so that practically all orders or families known appear suddenly and without any apparent transitions.[34]

"And he added:

> At the end of such a series, a kind of evolutionary running wild frequently is observed. Giant forms appear, and odd or pathological types of different kinds precede the extinction of such a line.[35]

"He concluded:

> within the slowly evolving series, like the famous horse series, the decisive steps are abrupt, without transition: for example, the choice of the middle finger for further transformation, as opposed to the two

---

[32] Mayr, Ernst. 1988. Toward a New Philosophy of Biology: Observations of an Evolutionist. Cambridge, MA: Harvard University Press, pp. 529-530.

[33] Goldschmidt, Richard B. 1952. Evolution, as Viewed by One Geneticist. American Scientist, p. 97, January.

[34] Goldschmidt, Richard B. 1952. Evolution, as Viewed by One Geneticist. American Scientist, p. 97, January.

[35] Goldschmidt, Richard B. 1952. Evolution, as Viewed by One Geneticist. American Scientist, p. 97, January.

middle fingers, in the evolution of the artiodactyls; or the sudden transition from the four-toed to the three-toed foot with predominance of the third ray.[36]

"This fact has also seeped into the popular press. The Cambrian explosion was documented in an article by journalist J. Madeleine Nash, who wrote: 'All but one of the phyla in the fossil record appeared' rapidly in geological time. She added that

> Zircon dating, which calculates a fossil's age by measuring the relative amounts of uranium and lead within the crystals, had been whittling away at the Cambrian for some time. By 1990, for example, new dates obtained from early Cambrian sites around the world were telescoping the start of biology's Big Bang.[37]

"Also, Nash noted that using the

> information based on the lead content of zircons from Siberia, virtually everyone agrees … 'We now know how fast "fast" is … And what I like to ask my biologist friends is, how fast can evolution get before they start feeling uncomfortable?'[38]

"In an article titled, 'The Evolution of Life on the Earth,' Harvard paleontologist Stephen Jay Gould[39] wrote, 'We do not know why the Cambrian explosion could establish all major anatomical designs so quickly.' The sole exception was 'Bryozoa, a group of sessile and colonial marine organisms, [which] do not arise until the beginning of the subsequent Ordovician period, but this apparent delay may be an artifact of failure to discover Cambrian representatives' adding:

> Although interesting and portentous events have occurred since [the Cambrian explosion], from the flowering of dinosaurs to the origin of human consciousness, we do not exaggerate greatly in stating that the subsequent history of animal life amounts to little more than variations on anatomical themes established during the Cambrian explosion.

"Furthermore:

---

[36] Goldschmidt, Richard B. 1952. Evolution, as Viewed by One Geneticist. *American Scientist*, p. 97, January.

[37] Nash, J. Madeleine. 1995. When Life Exploded. *Time* Magazine, p. 74, December 4.

[38] Nash, J. Madeleine. 1995. When Life Exploded. *Time* Magazine, p. 74, December 4.

[39] Gould, Stephen Jay. 1994. The Evolution of Life on the Earth. *Scientific American*, p. 89, October.

The sudden appearance of the flowering plants (angiosperms) was, to Darwin, 'an abominable mystery.' In his words: 'The rapid development, as far as we can judge, of all the higher plants within recent geological times is an abominable mystery.'[40]

"Niles Eldredge, who as you know is the chairman and curator of invertebrates, American Museum of Natural History, wrote that at

the core of punctuated equilibria lies in an empirical observation: once evolved, species tend to remain remarkably stable, recognizable entities for millions of years. The observation is by no means new: nearly every paleontologist who reviewed Darwin's *Origin of Species* pointed to his evasion of this salient feature of the fossil record. But stasis was conveniently dropped as a feature of life's history to be reckoned with in evolutionary biology. And stasis had continued to be ignored until Gould and I showed that such stability is a real aspect of life's history that must be confronted.[41]

"He added that in order 'to establish the plausibility of the very idea of evolution, Darwin felt that he had to undermine the older (and ultimately the biblically based) doctrine of species fixity. Stasis, to Darwin, was an ugly inconvenience.'"

I continued, "'For millions of years species remain unchanged in the fossil record,' said Stephen Jay Gould of Harvard, 'and they then abruptly disappear, to be replaced by something that is substantially different but clearly related.'[42]

"Harvard University professor of geology, Peter Williamson, wrote in *Nature*[43] 'all lineages exhibit morphological stasis for very long periods of time' and Neo-Darwinism 'has failed to predict the widespread long-term morphological stasis.' Williamson also wrote in his book that the principal problem in evolution

is morphological stasis. A theory is only as good as its predictions, and conventional Neo-Darwinism, which claims to be a comprehensive explanation of the evolutionary process, has failed to predict the

---

[40] Darwin, letter to J.D. Hooker, July 22, 1879.

[41] Eldredge, Niles. 1985. *Time Frames: The Rethinking of Darwinian Evolution and the Theory of Punctuated Equilibria*. Portsmouth, NH: Heinemann, pp.188-189.

[42] Biochemist and former editor of *New Scientist,* science writer Roger Lewin, wrote in an article titled, "Evolutionary Theory Under Fire" that "A historic conference in Chicago challenges the four-decade long dominance of the Modern Synthesis" in *Science* 210:883-888, 21 November 1980.

[43] Williamson, Peter. 1981. Morphological stasis and developmental constraint: Real problems for Neo-Darwinism. *Nature* 294:214-215, November 19.

widespread long-term morphological stasis now recognized as one of the most striking aspects of the fossil record.[44]

"Dr. Robert G. Wesson, in his MIT Press book, *Beyond Natural Selection,* wrote that the

> nearly timeless species are not exempt from the changes of proteins that go on in all living beings, and they could surely vary in many ways without loss of adaptiveness, but [they don't, and] their patterns have become somehow frozen. ... From the point of view of conventional evolutionary theory long-term stasis is hard to explain. Rapid evolution is comprehensive as species adapt to new conditions or opportunities, but it is incongruous that species remain unchanged through changing conditions over many millions [of years].[45]

"The bony-finned Coelacanth, long thought to be extinct eons ago, was rediscovered in 1938. It is called a living fossil because Darwinists believe it has remained the same for some 450 million years. Darwinists do not have a good explanation for the Coelacanth living fossil. Zoologist and professor for the public understanding of science at Oxford University, Richard Dawkins, in his book *The Blind Watchmaker*, wrote[46] that more

> biologists agree that stasis is a real phenomenon than agree about the causes. So, we have stasis … How do we explain it? Some of us would say that the lineage leading to Latimeria [the modern coelacanth] stood still because natural selection did not move it. In a sense, it had no "need" to evolve because these animals had found a successful way of life deep in the sea where conditions did not change much. Perhaps they never participated in any arms races.

"He added that:

> Their cousins that emerged onto the land did evolve because natural selection, under a variety of hostile conditions, including arms races, forced them to. Other biologists … say that the lineage leading to modern Latimeria actively resisted change, in spite of what natural selection pressures there might have been. … It is conceivable that coelacanths stopped evolving because they stopped mutating—

---

[44] Williamson, Peter. 1981. Morphological stasis and developmental constraint: Real problems for Neo-Darwinism. *Nature* 294:214-215, November 19.

[45] Wesson, Robert G. 1994. *Beyond Natural Selection.* Cambridge, MA: MIT Press, pp. 207-208.

[46] Richard Dawkins, in his book *The Blind Watchmaker* (1986), wrote this on pp. 246-247.

perhaps because they were protected from cosmic rays at the bottom of the sea!

"Both Gould and Eldridge attempted to explain away the serious fossil record problem by claiming that evolution occurs so rapidly that it leaves little or no fossil record, then stasis dominates for eons. Or, some claim, the transformation species population was so small that little chance existed for its fossilization. They called this new view *punctuated equilibrium*. Furthermore, even if true, this hardly explains the *virtually complete absence* of all of the critical transitional forms.

"The major problem is that this creative theory is *not* based on evidence, but rather on a *lack* of evidence. This pathetic excuse to explain away the evidence is like claiming that life exists, or once existed, on Mars; we just need to keep looking for the evidence. It's there; we just need to come up with more sophisticated techniques to find evidence for past life. We have spent billions of dollars using sophisticated techniques to find evidence of life on Mars, but so far, no clear evidence of life has emerged. None. Maybe the truth is, life is not there, and we are wasting precious resources looking for it that we could be using to cure diseases such as cancer."

After the debate, I asked Dr. Rushfield why he just walked out before we had hardly begun the debate. His only response was, "You used lots of quotes, likely all out of context, a tactic called quote mining, and I could not look each one up to show that they were all out of context, as I am sure they all were."

I responded to this claim by noting, "If I would have given the same information without referring to an authority, you would claim that I was just plain wrong or ask, 'What are your sources for that claim?' And I will be glad to give you the reference to each quote so that you can look them up. Then we can have a follow-up debate."

"No thanks. I do not have time and do not want to waste my time by looking up a lot of quotes. I am sure they are all out of context."

"It seems that you cannot win in these debates, no matter what you say," I answered at this point, feeling frustrated yet also resigned.

Dr. Rushfield then added, "Besides that, these debates give you credibility, as if your side had some validity, when it does not. I never should have agreed to be involved. I realized it was a mistake after ten minutes into the so-called debate. Never again will I be involved in debating a creationist. That's the last time."

Chapter 13: The Second Debate

Not long after the ruckus from this debate had quieted down, I was sent a letter by Professor Charles Hillmacker from Harvard. He wanted to debate me! Word got back to me that he was enormously embarrassed by the performance of the last person I debated. He had written several books against creationism and Intelligent Design, so he felt confident that he could debate me, so I agreed to debate him. We agreed on a date, and when the time came, we assembled in the college auditorium. I was very nervous when I heard his introduction, which was:

Introduction by the department chair: Dr. Hillmacker is professor emeritus of anatomy and an expert on the human body, specifically the kidney. We decided that rather than a debate, he would state his claim that demonstrated creation was wrong and evolution was correct. (It was soon clear that Hillmacker focused almost totally on the now fully refuted poor-design argument.)

(I then had ten minutes to relate my background. After this, the first argument Hilmacker made was that Darwin's claim of useless organs in humans proved his [Darwin's] theory or at least was a major argument for the theory.)

Hillmacker: The greatest biologist that ever lived, Charles Darwin, whose teachings were published in 1859, proved many organs to be useless, or have less function than they had in our evolutionary ancestors. The thyroid gland is one organ Darwin mentioned to prove that we humans evolved and were not created. Why would some god design organs that have no function? Evolution explains this fact quite well. This organ once had a function, but it lost that function as we evolved. The thyroid must be useless because surgeons found that adults could survive after having their thyroid removed. It also shrinks as we age and is replaced by fat. There, I have proved my case!

(It was immediately apparent to me that Dr. Hillmacker was grossly out of date.)

Radinow: The thyroid is a bilobed gland connected by a narrow isthmus located just inferior to the larynx. The thyroid is one of the largest endocrine glands and can grow to as large as twenty grams in adults. Its three most important hormones are triiodothyronine (T_3), thyroxine (T_4), both metabolism regulators, and calcitonin, which regulates calcium levels. Both T_3 and T_4 stimulate the mitochondria to provide more energy for the body and increase protein synthesis. Without T_3 and T_4, humans become sluggish, and growth stops. An oversupply (or an undersupply) of thyroxine results in overactivity (or underactivity) of many organs. Defects in this organ at birth can cause a hideous deformity known as cretinism that results in severe retardation of both physical and mental development.

Hillmacker: How do you know this? I admit I have not read the literature in a long time, as my specialty is the kidney, but your claim just seems wrong.

Radinow: My answer is that it has a clear, documented function. Dr. Kocher attempted to contact his 102 thyroidectomy patients. Of the seventy-seven patients he was able to contact, he found that of the twenty-eight patients in whom he had performed a partial gland removal, all were in good health. In contrast, of the twenty-four in whom the gland had been completely ablated, twenty-two showed clear deterioration.[47] Kocher was awarded a Nobel Prize in 1909 for his work in this area. The thyroid gland is now known to secrete several hormones essential to normal body growth in both infancy and childhood.

(I was astounded at how antiquated his knowledge was, but then this was not his area of specialty. He was a worldwide expert on the kidney, but he had retired almost ten years before.)

Hillmacker: Well, you may have answered this one, but what about the uvula. Surely it is a useless organ, if there ever was one. All it does is get in the way and cause sleep apnea and then it has to be removed.

Radinow: Not really. The most well-documented function of the uvula, in tandem with the back of the tongue, the palate, and the lungs, is to play a role in the articulation of guttural and other human voice sounds.[48] Uvular consonants are rarely used in English, but they are commonly used in many languages. These include Spanish, German, French, and Portuguese, plus a few Celtic languages, as well as some Danish and Scandinavian dialects, and many Semitic, Caucasian, and Turkish languages. The uvula also allows singers to produce a vibrato sound, from a wavy soprano to a bass voice. It also lowers the resonance of the air column over the larynx to reduce the extremes of voice nasality.

(Dr. Hillmacker was at this point shaking his head and saying nothing, so I continued.)

---

[47] Tröhler, Ulrich. 2010. "The Subtle Knife." *Karger Gazette* 71:12-14, pp. 12-14, October; Tröhler, Ulrich. 2011. Towards Endocrinology: Theodor Kocher's 1883 Account of the Unexpected Effects of Total Ablation of the Thyroid. *J R Soc Med* 104:129-132.

[48] Finkelstein, et al., 1992, Otolaryngology Head & Neck Surgery 107(3):440-450. p. 448.

The uvula is unique to humans, and no comparable organization of serous and mucous glands is found in the homologous velum of other mammals studied. This would be logical if its role is primarily as a speech organ. One study concluded that the uvula may be yet "another structure that differentiates man from other mammals.

In humans, it varies greatly in size and shape, from a tiny knob to a size reaching almost to the opposite side of the throat.[49] The variability of the uvula may help to explain voice distinctions, which help us to differentiate people on the basis of their speech.

The uvula also produces seromucous fluid to provide the proper lubrication for complicated human speech, as illustrated by the dry throat problem that occurs in some speakers before addressing a large audience. Studies of patients lacking the uvula have concluded that it has an abundance of seromucous glands which can produce a large amount of saliva in a very short period of time. The fact that the uvula can produce and secrete large quantities of thin saliva is confirmed by a common complication of surgery to remove the uvula, pharyngeal dryness. During phonation and swallowing, the uvula swings back and forth in the oropharynx, thereby basting the throat, helping to keep it moist and well lubricated.

For this reason, Finkelstein et al. concluded that a "major function of the uvula" is as a lubrication organ. Talking causes intermittent opening and closing of the velopharyngeal valve, which provides continued lubrication, critical for normal speech. Removal of the uvula usually results in a dry throat problem, which consequently causes hoarseness. Thus, the evidence from numerous studies is clear: "the uvula plays a very important role in moistening the oropharyngeal mucosa. The uvula also has large drainage canals that help to drain excess saliva from the oral cavity towards the base of the tongue.

Hillmacker: These are very minor functions and do not help survival (he answered smugly).

Radinow: I never said it was critical to survival, just that it had an important use! If critical to survival was the criterion, a lot of structures could be claimed to be vestigial, and the concept becomes meaningless.

Hillmacker: OK, what about the *nictitating membrane* remnant in the human eye? It is totally useless in humans but critical in some animals. The nictitating

---

[49] Finkelstein, et al., 1992, Otolaryngology Head & Neck Surgery 107(3):440-450. p. 448.

membrane, often called a "third eyelid," is a very thin muscle-controlled transparent membrane that moves horizontally across the eye surface to clean, moisten, and protect the eye. The membrane is hinged at the inner side of the lower eyelid of many animals. The nictitating membrane is especially important in animals inhabiting certain environments. This includes living near the ground and thus exposure to dust and dirt, as in the case of birds, reptiles, and mammals, or marine animals such as fish.

Radinow: The classic eye anatomy textbook by Snell and Lemp accurately describes the misnamed "nictitating membrane" that we now recognize as a semilunar fold of the conjunctiva. The plica is fully functional in humans. The human plica serves both as a support and control structure for the eye, as well as for lubrication to ensure effective eyeball movement. Furthermore, "the fold intercepts foreign bodies on the cornea and passes them to the region of the lacrimal caruncle, the part of the eye closest to the nose.

The plica significantly increases the field of vision possible without moving the head."[50] The eye has about 50 to 55 percent rotation, but without the plica semilunaris, the rotation would be considerably reduced. The reason is that it takes up the slack, which occurs when the eye looks forward or medially. No such arrangement exists laterally because the fornix in this area is very deep. Because the plica allows generous eye rotation, it actually is an example of over-design.

Hillmacker: What about the tail, called the coccyx, which is what is left of the tail we inherited from our ape common ancestor? I got you there!

Radinow: Not really. The major function of the coccyx is an attachment site for the muscle fibers and tissues supporting the bladder neck, urethra, uterus, rectum, and a set of structures forming a bowl-shaped muscular floor, collectively called the pelvic diaphragm. Without the coccyx and its attached muscle system, humans would need a very different support system for their internal organs, requiring numerous design changes in the human posterior.

Hillmacker: I cannot respond to all this talk except to note that evolution is the consensus view of scientists. All scientists accept it! It is consensus science.

---

[50] King, B. G. (1926). The influence of repeated rotations on decerebrate and on blinded squabs. Journal of Comparative Psychology, 6(6), 399–421. https://doi.org/10.1037/h0071155

Radinow:	The problem with consensus science is that the courts have ruled that opposition to evolution is teaching backdoor creation. Thus, most people will not learn about the evidence for the other side. Furthermore, few people on their own read the other side. How many of you have read books or journal articles supporting the other side? Let's see a show of hands.

(No one raised their hand.)

See, I told you so. None of you are aware of the other side.

Hillmacker: The reason is we know books supporting Intelligent Design are a bunch of crap written by ignorant people, and we do not want to waste our time.

Radinow:	Do you think I am an ignorant person?

Hillmacker: We are still attempting to figure out your problem.

Radinow:	How can you condemn books you have never read?

Hillmacker: We don't need to read books written by ignorant bigots. Evolution is consensus science, and only ignorant fundamentalist people read the other side.

Radinow:	Consensus is only an opinion, based on what a popular vote finds, but the facts of anatomy are clear. It is also based on people who have not read material on the other side. A major functionless organ, the tonsils, is the classic example. I had mine out as a young man, as did most of my friends. And the reason, in most cases, was for prophylactic reasons, better to get them out when you are young rather than wait until you are older when more problems are common. Their only function, as we used to say, was to help the doctor pay for his new Mercedes! However, now they are not removed unless clear medical criteria exist justifying it.

Hillmacker: I can agree with that! I have removed a ton of them in my day as a surgeon.

Radinow:	Now we know that the tonsils and adenoids are made up of lymphoid tissue that manufactures antibodies against invading diseases. The tonsils are critical in helping to establish the body's defense mechanism that produces disease-fighting antibodies. The tonsils are larger in children than in adults, because they are important in the development of the entire immunological system. Once these defense mechanisms are developed, the tonsils begin to shrink in the preteen years to almost nothing in adults, and other organs take over this function. The location of the tonsils and adenoids allows them to act as a trap and first line of defense against inhaled or ingested bacteria and viruses.

Therefore, unless there is an important and specific reason to have them removed, it is better to leave the tonsils and adenoids in place. The tonsils are continually exposed to the bacteria-laden air we breathe and, for this reason, can readily become infected. As part of the body's lymphatic system, they function to fight disease organisms. The tonsils form a ring of lymphoid tissue guarding the entrance of the alimentary and respiratory tracts from bacterial invasion. Called "super lymph nodes," they provide a first line of defense to bacteria and viruses that cause both sore throats and colds.

Hillmacker: Well, I agree with what you said, but they are not really necessary. You can do fine without them.

Radinow: If you really believe that, let's do a scientific study. Select a thousand healthy young people from four to six years old and randomly divide the sample in half. Half will have no treatment except a fake tonsillectomy, and the other half will have an actual tonsillectomy. Do not tell the parents which child had the real tonsillectomy. Then compare the two groups after two, four, and six years and see how they do. I bet you will find a difference in health between the two groups, namely that the tonsillectomy group as a whole will do worse. And I bet this experiment will never be done because we all know that at some level it would be a major failure and probably end up the subject of big lawsuits.

Hillmacker: I will give one clear, sure example of a worse than useless organ: the vermiform appendix. As my surgeon friends used to say, its only function is to help the surgeon pay for his second home.

Radinow: So that we are not here all night, I will just list a few of the lately discovered important roles for the appendix. The first is the safe house role.

Hillmacker: The what?

Radinow: Let me explain. Most bacteria in a healthy human are beneficial and serve several functions, such as helping to digest food and produce vitamins. If the intestinal bacteria are purged, such as by antibiotics or diseases such as cholera or amoebic dysentery, one of the appendix's functions is to reboot the digestive system with beneficial bacteria. Its location—just below the normal one-way flow of food and germs in the large intestine in a sort of gut cul-de-sac—supports the safe house role by protecting and fostering the growth of the "good germs" needed for various uses in the intestines. Diarrhea is designed to flush out all bacteria from the colon, both good and bad. The design of the appendix prevents it from being flushed by diarrhea, so that bacteria in the appendix are not affected by diarrhea and can rapidly repopulate the colon.

The appendix is also involved in the production of molecules aiding in directing the movement of lymphocytes to other body locations. During the early years of development, the appendix functions as a lymphoid organ, assisting with the maturation of B lymphocytes and in the production of immunoglobulin A (IgA) antibodies. Lymphoid tissue begins to accumulate in the appendix soon after birth and reaches a peak between the second and third decades of life. It decreases rapidly thereafter, practically disappearing after the age of about sixty.

The appendix also functions to expose white blood cells to the wide variety of antigens normally present in the gastrointestinal tract. Thus, like the thymus, the appendix helps to suppress potentially destructive humoral (blood- and lymph-borne) antibody responses while concurrently promoting local immunity.[51]

Hillmacker: I can't dispute you here either, but surely the thymus is an example of a useless organ. The thymus is a small pinkish-gray gland located below the larynx and behind the sternum. After all, it shrinks to nothing as we grow older, so it must not have a use.

Radinow: Wrong again! A capsule, from which trabeculae extend inward, surrounds it, dividing it into several lobules, each of which contains functional units called follicles. The thymus gland is an example of an important organ long judged not only vestigial, but harmful if it becomes enlarged. For generations, physicians regarded it as a useless, vestigial organ. An oversized thymus was once routinely irradiated to shrink it. However, it was later found from follow-up studies that instead of helping the patient, such radiation treatment resulted in abnormal growth and a *higher* level of infectious diseases persisting longer than normal.

This once-labeled "vestigial" structure is now known to be the lymphatic master gland. Without it, the T cells, which protect the body from infection, could not function properly because they develop within the thymus gland. Far from being useless, the thymus regulates the intricate immune system that protects us against infectious diseases. Thanks to these discoveries, scores of researchers are now pursuing new

---

[51] Thapa, P. 2019. The Role of the Thymus in the Immune system. Thorac Surg Clin . 2019 May ; 29(2): 123–131. doi:10.1016/j.thorsurg.2018.12.001.

and highly promising lines of attack against a wide range of major diseases, from arthritis to cancer.

The cortex of the thymus is densely packed with small lymphocytes surrounded by epithelial-reticular cells. The lymphocytes, also called thymic cells, are produced in the cortex and exit the gland by way of the medulla. The medulla is more vascular than the cortex, and its epithelial-reticular cells outnumber the lymphocytes.

Yet other functions of the thymus include both regulation and a dominant role in reducing autoimmune problems where the immune system attacks the person's own cells. Evidence now exists that regulatory cells have a role in preventing reactions against self-antigens, a function as important as their role of clonal deletion of high-affinity self-reactive T-cells.

Regulatory T-cells also help to prevent inappropriate inflammatory responses to nonpathogenic foreign antigens. This system plays an essential role in preventing injurious inflammatory responses to innocuous foreign antigens coming in contact with mucosal surfaces, such as in many allergies. A primary function of the thymus is facilitating maturation of the small white blood cells called lymphocytes, which are then sent to the spleen and the lymph nodes, where they multiply.

Hillmacker: You reject the science of Darwinism, so how do you know all of this? I thought creationists were uneducated true believers who got their conclusions from the Bible. And I am beginning to wonder about your information. You may be correct about the appendix, but it seems you are now misquoting authorities. You creationists have a reputation for doing that. I will be charitable and go on. The pineal gland surely is an example of a useless organ. I remember that from college.

Radinow: Well, when you went to college, that was what was taught. The pineal gland is attached to the midbrain by a median stalk. Its functions include having a prominent role in the vital hormone-producing endocrine system.

The pineal gland is now also known to have a critical role in reproduction. It has long been known that a reduction in the amount of light reaching the eyes stimulates this small gland to synthesize and secrete an anti-gonadotrophic hormone, which results in marked attenuation of virtually all aspects of reproductive physiology. The pineal gland is a true neuroendocrine transducer and, as such, is able to convert, through rather complex neuronal circuitry, a photic-neural input into an anti-gonadotropic hormone output. The pineal gland is a very active member of the body's network of endocrine glands, especially during certain growth stages.

The pineal gland's most commonly mentioned function is its role in producing the hormone *melatonin*. Cells in the pineal gland produce melatonin from serotonin by the catalytic action of hydroxyindole-o-methyl transferase. Melatonin is produced mainly in the pineal gland of vertebrates but is also produced in a variety of other tissues.

Light-dark levels communicated to the brain from the retina to the pineal gland function to regulate melatonin levels. Melatonin is also a sleep-inducing hormone that depresses mood and alertness. This is why darkness is generally conducive to sleep.

The pineal gland is the primary controller of timing the onset of puberty, a critical developmental function. Melatonin regulates the production of anti-gonadotropin hormones that play a role in the retardation of gonadal development by blocking the effects of gonadotropic hormones. Damage to the pineal gland leads to precocious puberty in males. Conversely, if the pineal gland is overactive, puberty is delayed.

Before the advent of modern artificial lighting, the number of hours humans spent in darkness was much greater. Today, bright lighting found in almost all homes and offices may be affecting our reproductive cycle. The onset of sexual maturity at an earlier age, and even the higher rate of multiple births, may be one result of this "light pollution," due to exposure to a large amount of light during most of one's waking hours.

Studies completed on "pre-electric" Eskimos support the conclusion that light and the pineal gland are important in reproduction. When it is dark for months at a time, Eskimo women stop producing ova altogether, and men become less sexually active. When daylight returns, both the women and the men resume their "normal" reproductive cycles.

Hillmacker: I have to admit that I am impressed with your knowledge, but it appears to me you are taking things out of context. I was expecting to debate the Bible, so I admit that I am not very prepared, but I have one example that you cannot refute. Male nipples surely are useless because males do not produce milk and cannot breastfeed a baby!

Radinow: But because they are very sensitive to touch, they are a major erotic
           organ. In this sexually liberated society, I am surprised that anyone
           would mention this!

With that, we each gave closing arguments and called it a night. We both had a
lot of questions sent our way. Following the debate, I heard "through the grapevine"
that Dr. Hillmacker felt that I had a basic knowledge of anatomy but misquoted a lot
of authorities. When I later pressed him for examples, he explained, "Look, I am
very busy. Look up the examples for yourself. Never again will I debate a creationist.
They are unethical, and all you did was misquote like crazy."

He never gave me a single example.

Chapter 14: My Doubts About Darwin Surface at Cornell

Soon, much discussion was occurring about Intelligent Design at Cornell, partly due to the two debates I was in, and partly due to the negative media coverage and a lecture Cornell's president gave. An article reporting on my lecture under the headline "Cornell President Condemns Teaching Intelligent Design as Science,"[52] is as follows:

> A national movement to have Intelligent Design taught in science classrooms is "very dangerous," Cornell University's interim president, Hunter R. Rawlings III, said after taking up the issue Friday in a speech. But Mr. Rawlings charged that colleges were not engaging enough in the [scientific] debate.
>
> Mr. Rawlings spoke to hundreds of faculty members and trustees during his state of the university address, which typically focuses on the college's accomplishments and business. Intelligent Design is a theory that says the universe is too complex to be the result of evolution and natural selection, proposing that a higher power is responsible. Proponents say that alternatives to evolution should be taught in classrooms. But Mr. Rawlings denounced Intelligent Design as a "religious belief masquerading as a secular idea."
>
> Mr. Rawlings added, "Right now, this issue is playing out in school districts, cities, counties, and states across the country." In citing a recent report by the Pew Research Center in Washington, Mr. Rawlings said that 42 percent of Americans believe that creationism should be taught instead of evolution. "This is above all a cultural issue, not a scientific one," Mr. Rawlings said.
>
> "The Origin of Species" of 1859, … [and] the debate had resurfaced because the country is politically and culturally polarized. Cornell's staff should speak out [against] any blurring of the lines between religion and science, he said.
>
> John G. West, a senior fellow at the Discovery Institute in Seattle, which is a leader in the Intelligent Design movement, said that he was concerned that Cornell's president was "fanning the flames of intolerance. A college president is in a unique position to create an

---

[52] By Michelle York, dated October 22, 2005.

atmosphere of free speech. If he's implying that faculty don't have the right to discuss ideas, I'm very concerned."

His speech was sure giving me problems. It apparently energized the faculty to go on a witch hunt to expose and punish any Darwin Doubters here at Cornell. The list of Darwin Doubters put out by the Discovery Institute in Seattle, Washington, was also brought to their attention. One biology faculty, Professor Neil Rhoton, a large man who could be very forceful in his opinions, happened to notice that my name was on it and confronted me in the hallway.

"What's this?" he asked in anger.

"I signed the list of Darwin Doubters because I agree with the statement. Have you read it?"

"No, I have not. Why should I?"

"Let me read it to you."

> "We are skeptical of claims for the ability of random mutation and natural selection to account for the complexity of life. Careful examination of the evidence for Darwinian theory should be encouraged."

"That is a bunch of bull crap if I ever heard any," Professor Rhoton replied, almost screaming.

"What specifically do you object to?" I asked, trying to calm him down.

"The whole wretched thing is deceptive. It is a fact that random mutation and natural selection account for the entire complexity of all life everywhere. Evolutionary naturalism is better supported than the law of gravity, and to question it is ingenuous to say the least. Evolution is not a theory, but a fact, and I have to wonder about the sanity of anyone who questions it."

"Well," I responded, "The first wife of our own faculty member Carl Sagan, National Academy of Sciences member biologist Lynn Margulis, wrote in answer to the question "What kind of evidence turned you against Neo-Darwinism?" the following:

> What I want to see is a good case for gradual change from one species to another in the field, in the laboratory, or in the fossil record—and preferably in all three. Darwin's big mystery was why there was no record at all before a specific point [dated to 542 million years ago by evolutionists], and then all of a sudden, in the fossil record, you get nearly all the major types of animals.[53]

She added:

---

[53] From an interview with Lynn Margulis printed in *Discover* Magazine in the 16 June 2011 issue. See footnote #15.

Paleontologists Niles Eldredge and Stephen Jay Gould studied lakes in East Africa and on Caribbean islands, looking for Darwin's gradual change from one species of trilobite or snail to another. What they found was lots of back-and-forth variation in the population and then— whoop—a whole new species. There is no gradualism in the fossil record. … "Punctuated equilibrium" was invented [in an attempt] to describe the discontinuity in the appearance of new species. … [T]hese discontinuities are real.

Furthermore, I added: "My professors at Harvard stressed that all scientific conclusions are tentative. Look how many so-called facts of science were overturned as research progressed. Berry Marshal proved that pathogenic bacteria cause most gastrointestinal ulcers, not stress or acidic foods, as once universally believed by the scientific and medical community.

"Many other ideas that were once very controversial, such as heliocentrism, plate tectonics, the germ theory of disease, the transmutation of base metals into gold, and even biological evolution, are now all mainstream," I added, smiling, feeling that I had made my point.

"Evolution is different. The evidence is so overwhelming that it is not to be questioned under any circumstances," Professor Rhoton replied, getting angrier by the second.

Somewhat chagrined at his dogmatic claim, at this point I responded, "Nothing in science is exempt from questioning; everything should be open to being questioned, no exceptions. Are you saying that evolution is the only exception?"

"Yes, that is exactly what I am saying! I can see this is getting nowhere and I refuse to discuss this with you any further. If you have evidence, then publish it in peer-reviewed articles, and no creationist has, so you have no evidence," Professor Rhoton responded, obviously now very unreasonable and getting irrational.

"First of all, I and others have published much evidence against Darwinism, and I would be pleased to give you copies."

"Well, I refuse to read any of your pseudoscience. Furthermore, I have always thought that a lot of bad science gets through peer review, so I am not impressed by that argument."

"Then that only proves that evolutionary naturalism has become dogma, and dogma is anathema to science," I responded, which I almost immediately realized was a mistake. He then raised his fist to strike me, and I rapidly backed off, not saying another word. I realized this issue could not be rationally discussed with many people, especially university professors.

After this, he walked out, so I decided to finish the debate on my own.

Back at My Office

Several days later Professor W. Roland Gonzales said to me, "I understand you doubt Darwinism. Is that true?"

I responded, "I hold all theories in science tentatively, including Darwinism, as any good scientist should."

"Oh, really? When you come up with an anti-gravity machine, let me know," he responded, slyly smiling. "And how is work on your perpetual motion machine progressing?" he added sarcastically.

"Have you ever read *The Structure of Scientific Revolutions*?" I responded.

"No, and I don't intend to if it was written by some lame brained creationist!"

"No. It is one of the most influential books ever written in science," I responded, somewhat surprised that he was not aware of this classic. But by this time, the lack of reading by my colleagues no longer surprised me. Most of my colleagues did little reading except in their narrow specialty. Conversely, I read on average a book a week, mostly on evolution or some area of cell biology.

"Well, the more you know, the stupider you become," he shot back.

"Will you listen to one of our own faculty?"

"Of course," he answered.

"Our own William B. Provine, quoting John Ender, wrote: 'Natural selection does not act on anything, nor does it select (for or against), force, maximize, create, modify, shape, operate, drive, favor, maintain, push, or adjust. Natural selection does nothing.'"[54]

"That's just quote mining!"

"And claiming that I am quote mining is a conversation stopper, a way of responding to the words I just quoted without actually answering the claims made. If you read Provine's whole statement, you will see that it is not taken out of context."

"No thanks. I have many other things to do. Besides, some of us regard Provine as somewhat of a cantankerous crank."

"You are still avoiding what he said," I responded.

"I have a lot of work to do," he replied curtly and walked away.

---

[54] From Provine, William. 2001. "Natural Selection in the Wild." In *The Origins of Theoretical Population Genetics*. Chicago, IL: The University of Chicago Press, p. 199.

Chapter 15: The Issue Comes to a Head at Cornell

Soon, my doubts about Darwinism came to a head in my department, and I was called in to see the Dean. I asked what it was about, but he said, "Just be there." I had a feeling that it was about the evolution issue and asked if I could bring a supporter and some of my published articles. He said, "No, just be there."

When I showed up, the dean was there, and two vice presidents, as well. They immediately pounded me with aggressive, hostile, rapid-fire questions. Actually, I felt that I was being interrogated, which, in fact, I was.

"I have had several student complaints about your teaching," the dean began.

"How many were there?" I asked, knowing that I always had excellent student evaluations, suspecting that this whole issue was related to my doubts about Darwinism.

"Three or four students from the Atheist Club," he answered. "But I cannot give you their names."

I then asked: "How can I respond if I do not know who they are?"

"We have to respect their privacy."

"What about my right to defend myself?" I asked.

"Sorry. I cannot give you any more information about the students."

"Could you give me more details about the complaints?" I asked.

"Yes, I can do that."

"Well, what was the problem?"

"From your lectures, they have assumed you do not believe in the standard orthodox theory of evolution," he responded.

I replied, "Well, you and I know that is a meaningless charge. I teach the evolution class here and have published extensively on evolution."

"I know that. They gave us evidence, such as a copy of a PowerPoint slide from Professor Ronald Number's book." The quote is as follows:

> In his revolutionary *Origin of Species* (1859), Darwin aimed primarily "to overthrow the dogma of separate creations" and extend the domain of natural law throughout the organic world. He succeeded spectacularly—not because of his clever theory of natural selection (which few biologists [then] thought sufficient to account for evolution) nor because of the voluminous evidence of organic development that he presented, but because, as one Christian reader bluntly put it, there was "literally nothing deserving the name of Science to put in its place." The American geologist William North Rice (1845–1928) … made much the same point. "The great strength of the Darwinian theory," he wrote in 1867, "lies in its coincidence with the general spirit and tendency of science. It is the aim of science

to narrow the domain of the supernatural, by bringing all phenomena within the scope of natural laws and secondary causes."[55]

"This claim is simply totally false. Who is this Ronald Numbers guy? … Obviously, he is an ignorant creationist," the dean added.

"He has a PhD in the history of science from the University of California, Berkeley, and is a professor of the history of science at the University of Wisconsin."

"I don't know about that, but I have been informed by several of your colleagues that you have concluded that there is no way molecules-to-man evolution could occur purely by natural means. As you know, this is heresy. How do you answer this serious charge?" he asked.

"The scientific literature is very clear. I will cite the medical biochemistry text, which we use, selected by the textbook evaluation committee, titled *Principles of Medical Biochemistry*[56] under the subtitle 'Mutations Are an Important Cause of Poor Health.' The authors wrote:

> At least one new mutation can be expected to occur in each round of cell division, even in cells with unimpaired DNA repair and in the absence of external mutagens. As a result, every child is born with an estimated 100 to 200 new mutations that were not present in the parents. Most of these mutations change only one or a few base pairs. … However, an estimated one or two new mutations are "mildly detrimental." This means they are not bad enough to cause a disease on their own, but they can impair physiological functions to some extent, and they can contribute to multifactorial diseases. Finally, about 1 per 50 infants is born with a diagnosable genetic condition that can be attributed to a single major mutation."

"They concluded:

> Children are, on average, a little sicker than their parents because they have new mutations on top of those inherited from their parents. This mutational load is kept in check by natural selection. In most traditional societies, almost half of all children used to die before they had a chance to reproduce. Investigators can only guess that those who died had, on average, more "mildly detrimental" mutations than those who survived."

---

[55] Numbers, Ronald. 2007 *Science and Christianity in Pulpit and Pew*. Oxford, UK: Oxford University Press, pp. 52-53, 279-280.

[56] Simmons, William H., and Gerhard Meisenberg, *Principles of Medical Biochemistry*. Amsterdam, Netherlands: Elsevier, p. 153.

I concluded: "It is clear that evolution is occurring, but it is going backwards, leading to genetic meltdown and eventual extinction. Over 99 percent of all mutations are near-neutral or harmful. The accumulation of mutations leads both to the aging of the individual and the species."

The dean responded, "Well, that textbook is wrong, and I have heard you say this before, but I don't buy it. Mutations selected as a result of the struggle for existence are causing the organism to constantly evolve upward to higher levels of complexity."

"That is not what the peer-reviewed scientific literature says," I responded.

He jumped back with, "I don't buy that at all."

"What percent of new mutations are harmful? 90 percent? 80 percent?" I asked.

"I have no idea," he shot back. "The simple fact is, we have other evidence of your teaching religion. One man published the following letter about you in the weekly *Cornell Post*. In case you have not seen it, here it is:

> As a father, I want my children to get the best education possible and to grow up in a world of opportunity. These goals are being threatened by right-wing fundamentalist creationists, including supporters of Intelligent Design, and by the typically left-wing "woo masters" with their crystals, chakras, and gross abuse of the concepts of quantum mechanics. Both are anti-science, and both pose a significant threat to the quality of education in the US. I feel obligated to get rid of these professors, a goal intended to help in some small way to counter their destructive influence on America, including one professor at Cornell who, of all things, teaches the evolution course.

"How do you respond?" he asked.

"He gave no evidence that what I was teaching was destructive or even wrong."

The dean then asked, "I hear that you have been reading books critical of Darwinism. Is that true?"

"Why are you concerned about what I have been reading? Isn't it my right as an American to read whatever books I want to?"

"Look, don't get smart with me," he threw back at me, obviously angry now.

"I have been reading a lot of books on both sides of the controversy for years now," I calmly answered. "And the science is very clear."

"Controversy? There is no controversy. None. Evolution from molecules to man is a fact as much as the Earth is round is a fact. Every scientist agrees with that."

"No, I know that thousands of scientists disagree with Darwinism, but not with microevolution."

As he raised his eyebrows, he asked, "What specific issue do they disagree with?"

"The disagreement depends on how you define evolution," I replied politely.

"Well, I never heard of any of these scientists," he noted.

I added that: "An article in *Life* magazine said, 'for all its acceptance as the great unifying principle of biology, Darwinism, after a century and a quarter, is in a surprising amount of trouble.'"[57]

"Look. *Life* magazine is not a scientific journal!" he shot back while narrowing his eyes.

"I do have an interest in the issue and have been thinking a lot about…"

"Well, stop thinking along the lines you are now thinking! It has gotten you into a lot of deep trouble. You know, your problem is that you think too much!"

"Are we open to new ideas at this university, or aren't we?"

"We are not open to pseudoscience, especially your anti-evolution pseudoscience."

"I invite you to read some of my papers to determine if my ideas are pseudoscience. All of them are well researched and fully referenced, usually with articles published in leading scientific journals including *Nature, Science, Cell,* and *The Proceedings of the National Academy of Science.*"

"I told you; I am not going to get into this! I will not waste my time reading any pseudoscience! Not yours or anybody else's. Is that clear?"

"How can you judge if my work is pseudoscience if you have not read any of my writings?"

"I do not need to read your writings to know it is pseudoscience or, much worse, religion."

"But, how can you know without reading them?"

"I know it is pseudoscience! I don't need to read it to know."

"I do not discuss creationism in my articles or classes, but I do cover the scientific problems with molecules-to-man evolutionary naturalism," I answered.

"Well, no Nobel laureate would agree with you!" he stated firmly. "There are no serious scientific problems with molecules-to-man evolution!"

"Not one?"

"Yes, not one."

"Actually, several do. For example, Arthur L. Schawlow, a 1981 Nobel laureate in physics, said, when confronted with the marvels of life and the universe, that one must ask why and not just how. The only possible answers are religious."

Ignoring what I just said, he pontificated: "The scientific consensus by qualified scientists is that evolutionary naturalism is a fact and any criticism of it is just back door creationism," he said again, as if repetition made his claim true.

He then interrupted himself, adding that "Creationism has been universally rejected by every scientist in the world."

"Except the 3,000 people on the list I noted earlier."

"I don't know about that."

---

[57] *Life* Magazine, April 1982, p. 46.

"Well, it is a list with real names in it, and information about where they teach or work, and their degrees. Would you like to see it?"

"No, I don't want to see it. I have better things to do with my time than look at some silly list."

"Science truth is not determined by the consensus of scientists, but by the scientific evidence and the scientific method, which I stress in every class that I teach," I noted.

"I agree with you there. I know that you do stress the scientific method in your classes. Nonetheless, evolution is accepted by every qualified scientist in the world. The scientific world is against you. The National Science Teachers Association, in their authoritative *Teaching Science in the 21st Century*, concluded that teaching doubts about molecules-to-man evolution 'is an insult to science education and an insult to the integrity of science.' Evolution denial, Professor Bybee says here on page 157, is 'grounded in a lack of understanding and ignorance.' Your doubts, as you call them, about molecules-to-man evolution is an insult to science grounded on ignorance. Can you deny this fact?"

"My research, both in the lab and the peer-reviewed literature, has found many major well-documented scientific problems with evolution. My list of almost 3,000 PhD-level scientists who agree that evolutionary naturalism from molecules to man has big problems proves this."

"This list does not surprise me. Even scientists can be crackpots, and many are. There are even a few here at Cornell."

"There are many well-known scientists, including a few Nobel laureates, on this list."

"Like who?

"Professor Ernst Chain, the founder of the antibiotics revolution. Dr. Wernher von Braun, the founder of our space program and the man responsible for putting a man on the Moon and the satellites in the sky that give us our modern telecommunications system, GPSs, weather satellites, satellite radio, and our modern internet that has revolutionized the world."

"I don't know about these so-called scientists, but, assuming your list is correct, my response is, only 3,000? I repeat, 99 percent of all scientists accept evolution as the explanation for all of life."

"True, but to question evolutionary naturalism has been ruled illegal in most state schools because it is viewed by the courts as back door creationism, and, in my experience, teaching it often ends one's teaching career, so what would you expect? Most scientists have never heard the other side of evolution. All they learn in college is the information in favor of the Darwinian worldview. So, I would expect most scientists would be Darwinists."

"I repeat, people who question evolutionary naturalism should be terminated. They have no business teaching our students pseudoscience and are dangerous. Evolution is the basis of all biology and, actually, all of science."

"I have heard this over and over *ad nauseum,* but it is not true, not here at Cornell or at the University of Michigan, Harvard, or Stanford either. I have reviewed the life science books used at these schools, and evolution is rarely mentioned in most textbooks except Introduction to Biology and Introduction to Zoology classes. I earned an MD and a PhD in cell biology from the most respected universities in the world, and evolution was rarely mentioned, even in graduate school. Even for my undergraduate classes, it was covered only in Introduction to Biology and Zoology."

"Well, that is a problem we have to fix. Here at Cornell, we are all going to stress evolution a lot more in our coursework in the sciences in the future."

"I am aware of that goal. At Harvard Medical School, the curriculum faculty committee is debating dropping the nutrition section and teaching a class on evolution instead. I told them that is irresponsible, as we know that poor nutrition is a major cause of illness and death, if not the major cause, in view of the obesity problem and its consequences, such as type II diabetes and heart disease."

"I think it is a very good idea to drop nutrition and stress evolution more. We need to improve our teaching of evolution. Students just do not know as much as they should about where we came from and where we are going, as evolution has documented."

"I have discussed this issue with a lot of people here and, in my experience, evolutionary naturalists actually not only know very little about evolution, but they also know almost nothing about the other side, and much of what they 'know' about the other side is wrong," I replied.

"I know you have talked about this topic with many others, and that is why you are in big trouble here."

"Questioning in science should not get anyone in trouble, ever. Scientists should question everything," I responded forcefully.

"True, except they should not question evolution, ever. Why do you do research in this area? You know it has gotten you into a lot of trouble."

"Because I find it very rewarding, to say the least. Every topic I have researched using peer-reviewed secular literature and my own research has supported the ID view and has falsified the Neo-Darwinian view. I have over ten years of research in this area now. All I ask you to do is review my research."

"I told you, I do not read articles on pseudoscience. This conversation is getting nowhere. I am going to strongly recommend to the faculty evaluation committee that you be terminated."

"Let me quote Greg Griffon, who has a PhD from our own university, who wrote that

> Biologists assume most of the variation seen in a given species is maladaptive and that only a small subset of a population attains optimality. But this is the opposite of what we see in nature. Variation

dominates in all species. If all traits were optimized by an ever-watchful natural selection, variation would quickly be eliminated in nature.

"An example he gave was:

Plant breeders have removed almost all of the variation in most of the fruits and vegetables we eat, which has resulted in predictable, market-ready (and often tasteless) food. But natural populations have none of the uniformity of conventional products. The moment a biologist opens the discussion to optimality, all of the seemingly maladaptive, anarchic, and random features of life stand as stark counterexamples.

"He added,

Biologists committed to the supremacy of natural selection seem to want to replace God the designer with Nature the designer. And, in that case, why can't Nature, in its wisdom, simply be a manifestation of God? In that regard, teleology (and its handmaiden optimality) plays right into the hands of intelligent design creationists."[58]

"Well, you will be hearing from me. That I can tell you."

A short time later, I explained to Professor Goldstean in the hallway, "I am amazed by the committee's response, but as the evidence against evolution mounts, and it is accumulating daily, science will look horribly bad in the future, just as the strong support for the Nazis by biology professors and medical doctors looks bad today. Of those at the Wannsee Conference where the Final Solution was decided upon, more than half of the participants were PhDs or MDs, and of those executed by the Nuremberg doctors' trial court, all were professors, lawyers, or doctors."

"The Nazi problem was that the leading Nazis were all good Christians," he confidently replied.

"This is not true. Hitler and the leading Nazis made it clear that after the war they were going to eliminate Christianity. In fact, the Nazis murdered more Christians than Jews, mostly Polish priests, and intellectuals."

"That's a bunch of malarkey if I ever heard it," he responded curtly.

I realized at this point that it was useless to discuss this history with him. He was badly misinformed. Badly misinformed. What I told him was extremely well-documented.

Then our German history professor, Dr. Helmut Meyer, who happened to overhear us, chimed in: "Dorland may be wrong about evolution, but he is correct about history. The scientists, mostly biologists and medical doctors, took the lead in

---

[58] Griffon, Gregory. 2004. *Evolution, Monism, Atheism, and the Naturalist Worldview*. Ithaca, NY: Polypterus Press.

the Nazi eugenic programs, including killing inferior races and deformed members of superior races."

With that, Professor Goldstean said nothing and quietly walked away. He obviously could not deal with being corrected. Dr. Meyer looked at me and said, as he walked toward his office, "I keep saying that we need to teach a lot more history, and this brief exchange verifies that conclusion."

Chapter 16: The Faculty Evaluation Committee Meeting

By the time I ended up before the faculty committee, I had over three hundred peer-reviewed scientific publications, the majority appearing in the leading scientific journals.

I was the sixth most highly cited biologist in the world out of a total of 628,000 whose research was cited at least once. The Science Citation Index reported that, as of that current month, my research had been cited over 47,212 times. Critical to my enormous success had been my students and co-workers, including forty-seven undergraduate researchers who had published papers with me, thirty-one successful PhD students, thirty-seven postdoctoral researchers, and eighteen visiting professors who had worked in my lab. I had published more extensively than any other professor at my university, and, as my peers knew, my record was so outstanding that I was recommended for early tenure by my department. It was also "undisputed" that I "covered the course material fully and was a well-regarded, highly successful teacher."

The meeting was on a Friday when most of us did not have classes to teach. The first question by Dean Simillae was, "Do you go to church?"

"Sometimes," I answered, somewhat chagrined over this objectionable line of questioning, adding, "You know this line of questioning is illegal. I have a law degree."

"We can cover our tracks. We have many highly paid attorneys, and we'll bankrupt you before you can prevail in court," Professor Simillae, smiling, curtly replied.

"Even if it is illegal to ask that question?"

"Yes. Creation is not religion but pseudoscience, so we are OK. Which church do you go to?" Professor Simillae asked again.

"I don't go to a church…"

"But you just admitted that you did!"

"I am Jewish and occasionally attend the local synagogue, not a church."

"Somehow, I doubt that. If you are a creationist, you must be a fundamentalist Christian. Are you a Baptist?"

"I told you I am Jewish and am not a creationist, but I do favor the Intelligent Design position."

"Intelligent Design is the same thing as creationism, as was proved in court in the Dover case."

"Science is not settled by court cases, but by scientific research," I added pensively.

"Well, that's your opinion, not mine."

Dean Cohen then added, "In the Dover case, it was proven that in a new edition of a book titled *Pandas and People,* that critically evaluated Darwinism, the term "creation" was changed to "Intelligent Design." Judge Jones ruled from this single example that Intelligent Design is just good old-fashioned creationism."

I responded, "You cannot judge a whole movement by the modifications made in a book to attract a wider audience. When the first edition of the book was published, the term "intelligent design" was largely unknown and rarely used. To change the word "Creation" to the phrase "Intelligent Design" was purely a marketing decision made by the editors. To determine what Intelligent Design supporters believe, you have to do a valid representative survey. This has never, ever been done, but it should be."

"And how do you know this?"

"I know from personal experience that the movement's followers hold a wide variety of views on evolution and creationism. Lastly, until ID came along, many Darwin Doubters had no group to associate with except creation groups, nor any place to publish except creation journals and books. Now that ID is around, a lot of Darwin Doubters have a home."

"That's nice, but both views believe that God had a role in creation, and science has concluded that the entire natural world and all of creation is the result of purely natural forces such as gravity. In science, no gods are allowed or even needed. Evolution can do it all. It can explain everything. Evolution created everything; the stars, planets, the Earth, and all life, including humans. The esteemed Professor Harold Morowitz wrote a book[59] titled, *How the World Became Complex: The Emergence of Everything.* It covers the evolution of everything and shows that everything evolved. No room exists for an intelligent designer to do anything. Evolution explains it all. Everything!" he added, to emphasize his point.

I then added, "I am also concerned about the effects of evolution. For example, in the United States alone from 1907 into the 1960s, approximately 30,000 people were sterilized for eugenic reasons. One study concluded that the

> Nazi sterilization law was applied much more radically than any other sterilization law in the world. They sterilized about 400,000 people within about 6 years before the 1939 decree to kill the disabled. The eugenic decree resulted in the death of about 200,000 people. About one out of every 200 people in Germany was sterilized.

"Well, that problem was due to creationists' adverse influence on science," he claimed.

---

[59] Morowitz, Harold. 2004. *How the World Became Complex:* The Emergence of Everything. New York, NY: Oxford University Press.

I was amazed at this claim! "It was clearly due to evolutionist teaching, not creationist influence. Creationists have very little, if any, influence in education today, especially due to court rulings."

"I see that we are not getting anywhere, so, after I confer with my colleagues, I will adjourn this meeting for now."

After that, I realized that he was a Darwinist true believer, and no amount of information would convince him. I then walked out of his office because I was very vexed at the endless irresponsible name-calling.

Chapter 17: The Secret Meeting

After my meeting with the dean, unbeknownst to me, the faculty had a private meeting, which I did not learn about until after my lawsuit was filed. It was recorded and transcribed, parts of which follow:

Professor Patricia Beckenworth: It is a shame Professor Rabinow came out as a critic of Darwinism. He was in line for a Nobel prize but, as an anti-evolutionist, it will now never be awarded.

Professor Herzig: He is not an anti-evolutionist. He has only concluded that evolution is going the wrong way due to the accumulation of near-neutral mutations.

Professor Goldstean: That claim is worse because he not only denies the fact that all life came about due to evolution, and mutations are our creator, but claims the opposite of evolution, namely that it is our destroyer.

Herzig: I have talked to him about this issue, and I believe he has some evidence for his view.

Goldstean: You know, defending him can be very problematic. I suggest you better watch it, or you will be next on the chopping block!

Herzig: Look, I only said *some* evidence. He is wrong; he has to be, as the idea that there is an intelligent creator is unacceptable to us. Science must be naturalistic.

Professor Rogers: Right. His idea that the accumulation of mutations causes an eventual mutational meltdown and extinction is religion, not science, because that idea supports the Genesis Fall in the Bible. And, as you noted, science must be naturalistic. Those who reject this fact will be blacklisted.

Beckenworth: Yes, we all know that. Let's get back to Professor Rabinow, shall we? He claims that the accumulation of mutations will eventually cause mutational meltdown and extinction.

Professor Rachel Child: Maybe we should respond to him with research, without invectiveness and name-calling as we are doing now.

Goldstean: He does not deserve a rational response. He is just plain nuts.

Professor Woodhouse: I agree with Dr. Herzig, but I have a question for Professor Beckenworth. How do you know that he will not be awarded the Nobel? Biochemist Dr. Kary Mullis, the inventor of the polymerase chain reaction, a billion-dollar industry now and a central tool for almost all DNA work, got a Nobel, and he is a certified heavy drug-using nut case. Have you read his book titled *Minefields*?

Herzig: I haven't. Has anyone here besides Professor Woodhouse? [No one answered.] It is obvious that no one here has. I would not call him a certified nut case, but only an eccentric, as are many great scientists. I would also be very careful about making drug use allegations about him. I respect him and his work greatly! I can tell you we would be out of business without PCR in my lab. It is critical for all of the biochemistry DNA research that we do.

Beckenworth: Let me explain how I know that Rabinow will never get a Nobel, no matter what he achieves in the area of science. I know because Professor Raymond Damadian, the inventor of magnetic resonance imaging, or MRI, was denied a Nobel due to the fact that he is a creationist. MRI is one of the most important technologies of the last century. We have four classes here based on MRI, and some medical schools offer even more classes on MRI than we do.

Woodhouse: How do you know he was denied a Nobel due to his rejection of evolution?

Beckenworth: Because a good friend of mine was on the committee. It was decided to exclude him from the award during the coffee sessions before the formal meetings.

Goldstean: But what evidence exists to prove he actually invented MRI? Sounds like a creationist myth to me. They are all infamous liars, you know.

Beckenworth: He proved it in court. General Electric, Siemens, Hitachi, Toshiba, Philips, and Shimadzu were all found by the court to have infringed on Damadian's patents and were forced to pay damages. They would not have been forced to do so if Damadian lacked proof. The damage award from General Electric alone was 128.7 million dollars, and Siemens, Hitachi, Philips, Shimadzu, and Toshiba all settled out of court for large, but undisclosed, amounts of money.

Goldstean: What was the basis of the court ruling?

Beckenworth: He has about 38 patents. I cover this in my classes and happen to have my notes in front of me. In March of 1972, Damadian filed for a patent for his MR scanner based on both his time 1 (T1) and time 2 (T2) discoveries and his voxel body-scanning method. Some specific landmark steps in MRI technology made by Damadian include:

1. Scientific research and theory of the cell led him to propose water molecule scanning by NMR as a method for detecting cancer.

2. Discovery of the cancer MR scanning signal in animal tissue, together with his demonstration of the pronounced diversity of MR relaxation times among healthy tissues.

3. Building a superconducting magnet (his competition at that time were all experimenting with commercially available electromagnets).

4. Filing of the original (and the foremost) patent on MR scanning.

5. Achievement of the first whole-body MRI scan of a human and the resultant image.

6. Development of the world's first commercial MR scanner.

Goldstean: I can see why they would deny him a Nobel. Denying evolution, the basis of all life sciences, is worse than being an active KKK supporter.

Woodhouse: Well, they are not as bad as the KKK nuts, but certainly close.

Goldstean: I think we have no choice but to take his graduate students away from him. Even if Dr. Rabinow is an excellent professor, mentoring under him now will handicap the careers of his students. We just cannot subject them to this problem. Also, if word gets out, which it already has if you read the newspaper and the internet posts about Rabinow, we have no choice now. Many of my colleagues are demanding that he be fired, and a few students have withdrawn from his classes due to this issue. Several students who posted on ratemyprofessor.com wrote that he is a creationist and, therefore, incompetent and a bigot. We may also have to not allow him to teach any more classes if this problem gets worse.

Beckenworth: But he is very well liked by his students and has high student ratings.

Goldstean: Yes, he does now, but once the word gets out to more students that he is a creationist or a Darwin skeptic or whatever you call him, this will likely change. I lecture strongly against Intelligent Design and all ideas against evolution in my classes, so the students know he is wrong.

Rogers: I move that we do a background check on every potential faculty hire to make sure he or she is an evolutionist and not a creationist or Darwin Doubter.

Goldstean: I agree. We must never hire a creationist. Not a single one. Jerry Coyne at the University of Chicago just wrote, and I quote, "I abhor discrimination against hiring simply because of someone's religion, but adherence to ID (which, after all, claims to be a nonreligious theory) should be absolute grounds for not hiring a science professor." I agree with Professor Jerry Coyne.

Herzig: But Rabinow was not a creationist when we hired him. His doubts about Darwinism came after about a decade of research here. How are we going to prevent this?

Goldstean: Simple: Each year, we just require all faculty to sign a sworn statement saying they are fully in agreement with evolutionary naturalism.

Herzig: Rabinow would have signed it when he was hired, so it would help if we required all faculty to sign it each year.

Woodhouse: No matter, we can't do that. This is what fundamentalist colleges do. We certainly do not want to emulate them.

Professor Harrington: Do you have a better idea? This problem is only going to get worse as the number of Darwin skeptics grows, which I am told is occurring now. Why don't we have them sign an agreement that if they become Darwin Doubters, they must resign?

Rogers: If we did that, they would just stay in the closet.

Goldstean: That does not help with the problem we have now. The problem now is Professor Rabinow, who will no doubt lose all of his grants and outside support. The anti-creation groups, such as Eugenie Scott's group, surely will write to his funding sources demanding that he no longer be funded. And, as an article in *Science* mentioned, he is an enemy of science, and no doubt his grant administrators will see this article, and that will be the end of his grant money.

Harrington: I am sure sooner or later they will find out that he has rejected the fundamental principle of all science, especially biology, namely evolution, and conclude that he should not be funded.

Goldstean: I think we need to keep our own list like the Darwin Doubters list, only it will not be public. That way, we can consult it when hiring faculty, and the granting agencies can also use it to ensure creationists do not get any research grants.

Woodhouse: I object to this approach. It is nothing short of blacklisting persons.

Goldstean: Do you have a better idea? We have to do something. We cannot let those who reject the most important principle of all science do research or teach here. Let him teach in a fundamentalist college.

Woodhouse: He is Jewish, and I doubt they would hire him.

Herzig: He is Jewish? No, no. I thought he was a Christian fundamentalist!

Woodhouse: We already discussed that at the start of the meeting! Weren't you paying attention? Yes, he is Jewish. Why would you think otherwise?

Herrington: Because he behaves like Christian fundamentalists do, you know, due to this creation business and all that.

Goldstean: I have sat here for over two hours and listened to everyone, and it is clear that he should be fired for just cause. He denies evolution and therefore rejects the basis of all science. Everything evolves, not just life, and evolution is the source of everything. The Big Bang started everything, then evolution took over and produced everything that is natural. The galaxy clusters evolved, the galaxies evolved, the stars evolved, the solar system evolved, the planets evolved, the geological formations evolved, the first life-forms evolved, the prokaryotes evolved, the fungi evolved, the plants evolved, the animals evolved, and lastly, humans evolved. No intelligent designer except humans was involved in creating any part of the universe that we see around us. No gods, angels, demons, spirits, or anything else. Evolution explains everything. As far as I am concerned, anyone who denies this fact should be fired. Gone. Out of here.

Professor Williams: But half of the American population believes that humans were created by God about 6,000 years ago, and 40 percent believe that humans were created by evolution with God's guidance. That leaves around 10 percent of Americans who agree with us. You would have to fire almost everybody. How can you deal with that?

Goldstean: That means as professors we have a lot of work to do. In all of my astronomy classes, I spend a major proportion of class time just trying to get across the fact that everything has evolved by natural means and there is nothing left for any god to do. We also study Steven Hawking's work, where he documents his M theory very well, proving God is irrelevant today in the age of science. Those students who do not accept this fact need to be warned of the fact that they may flunk. I do pre- and post-surveys, and from this data, I know that I am very successful in converting students. I am proud to say that I have converted many theistic students to scientific atheism.

Williams: I often wondered why you have so many students that drop your class, as well as so many students that flunk your classes.

Goldstean: Look, let's get back to Professor Rabinow. I have checked his résumé, all of his 300 publications, and found a few problems. He claimed to have a chapter in the Kalvert text, and I checked to verify that it was there. I found the chapter he claimed to have authored, and it was written by someone else. Here is a copy of his résumé and the book he claims to have written the chapter in.

Woodhouse: This only confirms our judgment. He is a liar as well! All creationists are liars. Anything more? If not, let us adjourn.

The Department's Position on Evolution and "Intelligent Design"

The department also decided to formally distance itself from Professor Rabinow, so they agreed to put on their department website the following disclaimer.

> The faculty in the Department of Biological Sciences at Cornell University is committed to the highest standards of scientific integrity and academic function. This commitment carries with it unwavering support for academic freedom and the free exchange of ideas. It also demands the utmost respect for the scientific method, integrity in the conduct of research, and recognition that the validity of any scientific model comes only as a result of rational hypothesis testing, sound experimentation, and findings that can be replicated by others.
>
> The department faculty, then, are unequivocal in their support of evolutionary theory, which has its roots in the seminal work of Charles Darwin and has been overwhelmingly supported by all scientific findings accumulated over the last 140 years. The sole dissenter from this position, Professor Dorland Rabinow, is a well-known proponent of "intelligent design." While we respect Prof. Rabinow's right to express his erroneous views, they are his alone and are in no way endorsed by anyone in the department. It is our collective position that intelligent design has no basis in science or fact, has not been tested experimentally, and should not be regarded as science but rather as religion.

This disclaimer passed the lawyers' scrutiny, was judged lawsuit-proof, and was posted on the university website.

The Court Case

To defend my academic freedom, I appealed to the courts. The US District Court, in a summary judgment, ruled against the university on most every count based on academic freedom, noting that faculty members are free to divulge their personal views in the classroom as long as they are not disruptive. The university appealed this decision to the circuit court, which agreed with the trial court's statement of facts, but the three-judge panel reversed the decision, thereby reinstating the ruling to censor me.

The court added, "restricting my speech was a part of" the university's right; and any handouts I used, even from major peer-reviewed scientific literature, or that mentioned my doubts about the sufficiency of Darwinian mechanisms to evolve all life from simple molecules purely by natural means could not be used in my classes, essentially preventing me from expressing the conclusions of my research on campus.

Chapter 18: The Aftermath

The university claimed, in writing, that it had the absolute right to restrict even "occasional in-class comments" and "even out-of-class lectures," mentioning "the professor's personal views on the subject of his academic expertise." I admitted that my "personal conclusions" colored my perspective on my subject, and I began each semester's biochemistry and cell biology classes with a brief discussion of my conclusions, namely, that Darwin's title of his 1871 book, *The Descent of Man* is indeed true, not the *Ascent of Man,* which is what he tried, but failed, to prove. We are descending, not ascending, or evolving, and evolutionary naturalism cannot produce us nor save us. I also occasionally mentioned my doubts about Darwinian mechanisms' ability to create the living world, quoting such scientists as our own Carl Sagan's first wife, Lynn Margulis. When the university forbade me from even mentioning any of this information *in class, I decided to go to court.*

I only learned about Professor Goldstein's false charges later and proved in court that my résumé was correct. The edition of the book in which I published my chapter was a newer edition than Professor Goldstein had at the meeting. The responsible thing to do would have been to ask me about my résumé entry, and not to make such irresponsible charges behind my back as he did. And this happened more than once. This meeting had been like allowing only one side to testify in a court trial.

In the end, after consultation with their attorneys, the faculty ruled that, as a tenured faculty member, I could not be fired based on my religion but could no longer be allowed to teach any biology classes. I was allowed to teach only the research design, biological statistics, and research methodology classes for fear of causing students to doubt Darwinism. To me, what they wanted to do was nothing less than pure indoctrination, not education, as we claim as a university, and nothing more.

My wife, Pam, amazed, asked, "Did the court actually accuse you of religious proselytizing?"

"Not directly, they ruled that allowing professors to present their own views in class implies the university endorses them, and it endorses 'everything it does not censor.' In my view, this claim was worse than ridiculous—most students know full well that professors disagree on topics related to politics and religion. My attorney argued that occasional well-thought-out conclusions at a public university 'cannot be construed as bearing the university's imprimatur and thus are protected under the First Amendment when they are non-disruptive and non-coercive,' a view that was rejected by the court. The fact is that my 'comments were non-disruptive, non-coercive, and were clearly identified as 'personal bias.'"

"This is absurd, absolutely absurd!" she exclaimed. "Why in the world did they rule this way?"

"The court ruled, in clear violation of my academic freedom, that the university had the absolute right to censor me, obviously only because the issue involved critiquing Darwinism. The university blocked me, and *only* me, from even briefly mentioning my personal views in the classroom, which I voiced to 'help students in understanding and evaluating' my classroom presentations. My attorneys also argued that if only those with an atheistic or agnostic evolutionary worldview could freely express their views, students might come to the erroneous conclusion that all professors shared that worldview."

The case was characterized by one human rights group as follows:

> The university administration ordered Dr. Rabinow to discontinue his classroom speech as well as his optional on-campus talks. No other faculty and no other topic have been similarly curtailed. Dr. Rabinow obtained a federal court order protecting his free speech and academic freedom, but it was overruled in a disastrous opinion by the US Court of Appeals.... The Court held that public university professors have no constitutional right of academic freedom and that their right of free speech in the lecture hall is subject to absolute control (censorship) by the University administration.

"Was that their only concern?" Pam asked.

"No, another concern was the university's use of derogatory labels to characterize my view of origins, referring to them as 'Bible belt' ideas and, therefore, 'inappropriate' at a university. The head of my academic unit even claimed that my beliefs 'hurt the reputation' of the university."

"Boy! What bigotry!" Pam blurted out.

"The *Amicus Curiae* and the brief prepared by my attorneys to appeal the case to the US Supreme Court documented the fact that the censorship I suffered is 'recurring on campuses throughout the nation.' They wrote:

> We are shocked at the breadth of speech rendered vulnerable by the court of appeals' decision.... [T]he decision... [gives] universities broad power to censor (Pet. App. A10). This view is completely antithetical to the premise underlying higher education—that students grow intellectually from confronting new or disturbing ideas, not from avoiding them....

"They added that the

> petitioner was reprimanded for his expressions solely because of the religious viewpoint presented in them.... [The university]... routinely permits faculty to present non-religious perspectives in the classroom

in their area of expertise. Dorland Rabinow's experience shows that such discrimination is, unfortunately, typical. Religiously committed academics in public universities across the country face resistance when they attempt, however briefly, to discuss or even disclose their ideological perspective in the course of their teaching or scholarship."

I then added, "In limiting my classroom speech, the court of appeals went far beyond limiting my speech only in the classroom. The court's rationale authorized limitations on other forms of my expression, such as:

> If a professor's in-class speech can be attributed to the university merely because he is employed there, so can his comments in the media or statements in his scholarly work. Surely, some students taking a professor's classes might feel discomfort or anger at his extra-curricular speech or scholarship.

"The amicus brief written in my defense noted,

> A major concern in this case is that the university imposed restrictions on Dr. Rabinow's speech that did not apply to any other professor and specifically singled out religious speech for censorship, but did not attempt to censor the religious speech of any other professor."

I added, "Pam, the appeals court held that the university could suppress religiously friendly speech merely to avoid a 'potential establishment conflict,' and even argued that the expression of a religious position in a secular subject, no matter how carefully presented, creates the appearance of endorsement of that position by the university and engenders anxiety in students who may feel compelled to feign a similar belief and, worse still, deny their own beliefs."

"That is appalling! Anxiety? I often faced anxiety from my far-leftist professors! I guess that's just fine!" Pam exclaimed.

I continued, "The circuit court appeal argued that the university restricted my speech 'solely because of its alleged religious content,' and argued that 'speech presenting a religious perspective is entitled to the same non-discriminatory treatment as other forms of speech, including anti-religious speech.' The court of appeals rejected this and authorized 'virtually limitless censorship of in-class or classroom-related speech by professors' if it can be construed as favorable to 'religious' or 'religiously motivated' views."

"Reminds me of totalitarian countries," my wife added.

"Yes, it does. Strictly applied, it would be inappropriate for a professor to state he is Jewish or Muslim, goes to church, or believes in God. Yet the same professor is allowed to state that he does *not* believe in God or does *not* believe in a theistic worldview. In brief, he can lecture *against* whatever the state defines as 'religious' values or beliefs, but not *for* them. As pointed out by law professor Phillip Johnson, the decision in this case reflects an obvious contempt for those who have serious questions about Darwinism, such as

> when Judge MaCardle explained why a professor of physiology was not allowed to tell his class about his doubts concerning the orthodox theory of human evolution... the university would certainly regard such a professor as an embarrassment and would try to keep the damage to a minimum.

"My attorneys argued that 'discomfort, anger, or anxiety on the part of a student or two cannot authorize suppression of a viewpoint' because the whole point of academic freedom is to protect speech *specifically* in cases where it might engender disputes, disagreements, discomfort, anger, or anxiety. Speech not generating these emotions is rarely suppressed, and thus its protection is of very little concern. *Anti-Christian speech* in universities clearly 'engenders anxiety' in Christian students, but efforts to suppress *that* type of speech have consistently failed."

My wife added, "If strictly applied, this would require courses in the history of Western civilization (and every other civilization) to expunge all discussion of the significant contributions religion and religious beliefs have made to civilization. Also, no college or university could offer *any* general course on philosophy, the history of philosophy, or even the history of science that included an honest discussion of the role of religion. Yet, the Court of Appeals ruled that the university *can* 'restrict speech'—even that which 'falls short of an establishment violation.'"

"Yes," I explained, "In brief, the court ruled that it can convict one of a First Amendment Establishment Clause violation, even if it rules the person's actions fell short of committing the violation! Courts traditionally have required *overwhelming evidence* that major negative effects have occurred, and not merely indications that such negative effects *might* have occurred, as the ruling in my case concluded. This decision signifies a new trend: for statements that can be interpreted as endorsing theism, all other considerations (including the First Amendment) can be ignored—any speech that may endorse theism can be suppressed."

The US Supreme Court rejected my *writ of certiorari* petition, and thus my case ended. A Petition for Writ of Certiorari is the document the losing party files with the Supreme Court requesting a review of the lower court decision. Rejecting it means that they refused to hear the case.

My case was critically important because, as Professor Johnson explains, the judges' opinion is a prime example of what he calls the "sham neutrality" of liberal rationalism, where "toleration, which may include the right to censor the 'insensitive' speech of others, is extended to some and denied to others without any explanation of the difference." The bias in this case was so extreme that a spokesman for *Americans United for Separation of Church and State,* Robert Boston, noted that "a Federal Court had applied the secondary-school ruling to a public university," negating the fact that courts view college students as "more mature and better able to judge" whether a professor's statements amount to institutional endorsement of religion.

"What was the reaction of your professional organization?" Pam asked as she sat down on the sofa.

"Robert M. O'Neil, general counsel for the *American Association of University Professors*, wrote that the judges had given university administrators far too much discretion, and the decision's wording was 'dangerous and very sweeping,' and "could represent an invitation for intrusion into the core of academic freedom—what goes on in the classroom.

"J. Scott Houser, executive director of the *Southern Center for Law and Ethics*, concluded that this appellate court decision should concern all faculty: 'In effect, it reduces the professor to a puppet of the university. The court held that the institution retains academic freedom, but professors do not.' Law Professor Johnson said of this case that the opinion by federal court of appeals Judge Floyd Gibson created the relevant principle that it is

> the right of educational administrators to control what is said in the classroom. 'The judiciary should not interfere with such internal university matters,' said Gibson, because 'federal judges should not be *ersatz* [substitute or replacement] deans or educators.' If Dr. Rabinow and other professors were dissatisfied with the restrictions placed on them by their academic superiors, their remedy was not to go to federal court but to seek employment at a different university that was more tolerant.

"This was like ruling that if a Black woman faced blatant discrimination, she should just look for a job elsewhere and not appeal to the courts. Wagner concluded that, 'viewed in the light of our tradition of academic freedom, the University failed in its responsibility to defend Professor Rabinow…all of us who teach in public colleges and universities …are left to contend with one of the worst judicial opinions on higher education in recent memory.' The courts have consistently ruled in *favor* of faculty who endeavor to inject anti-religious, atheistic, or agnostic material into their classes. One of the most extensive studies on academic freedom was completed by a Columbia University committee. On the important question 'Does academic freedom mean the freedom of the academy or the freedom of the scholar in the academy?' the committee ruled:

> intellectual life consists in the activities of a faculty, including, in the first place, their relation to the students. It is educational freedom that is at issue. *The academy is free when the scholars who make it are free, as scholars.* And the academy is free when its governing board is free to protect and to advance this freedom.

"In my case, *the* court clearly violated this widely accepted standard. Academia is not free when the educators who are in the academy are not free.

"A result of these court cases has been that both the persecution and ghettoization of Darwin Doubters have increased significantly. For example, the president of the University of Idaho, Dr. T.P. White, instituted a campus-wide classroom speech code on October 4, ruling that 'evolution' was 'the only view that is appropriate' for science classes. My attorney added that this decision goes along with the trend in education.

"The University of California at San Diego's website stated that 'all first quarter freshmen' were 'required to attend' a lecture at the campus's sports arena given by an anti-ID activist, titled, 'Why the Judge Ruled Intelligent Design Creationism Out of Science.'

"Biology professors at Southern Methodist University taught a course attacking ID. The course website stated, 'You don't have to teach both sides of a debate if one side is a load of crap.' A leading evolutionary biologist at the University of Chicago, Jerry Coyne, stated that 'adherence to I.D. should be absolute grounds for not hiring a science professor.'

"A professor of biochemistry and leading biochemistry textbook author at the University of Toronto wrote that a major public research university 'should never have admitted' students who support ID and should 'just flunk the lot of them and make room for smart students.'

"Three leading biology professors at Ohio State University derailed a doctoral student's thesis defense by writing a letter claiming that 'there are no valid scientific data challenging macroevolution' and therefore, the student's teaching about problems with Neo-Darwinism was 'unethical' and 'deliberate miseducation.'

"A Biology 101 lecturer at Wesleyan College endorsed teaching students 'inaccuracies' that are 'wrong' if that enables educators to 'gain their trust' and 'help them accept evolution.' In other words, 'Lying in the Name of Indoctrination into Darwinism,' is just fine.

"In an article 'Want a Good Grade in Alison Campbell's College Biology Course? Don't Endorse Intelligent Design', a biology professor at the University of Waikato stated, 'If, for example, a student were to use examples such as the bacterial flagellum to advance an ID view, then they should expect their grade will be lowered.'"

"Sounds worse than the Inquisition!" Pam said.

Sadly, in some ways it was.

Chapter 19: A New Chapter in My Life

After I lost my court case, I lost all my graduate students and my research funding and was removed from all faculty committees. As my activities were now very limited, I taught the few classes they would allow me to teach and spent much more time reading. As I read, I became increasingly convinced of my conclusion that evolutionism has been falsified by valid scientific research. Only a few months after I lost my court case, much of the chimp genome was sequenced.

A review in *Nature* and other scientific journals covering the analysis completed by several labs concluded, "it is clear a chasm exists between the human and chimp genomes. When the analysis of all of the comparisons was completed, including pericentric inversions, chromosomal fusions, deletions, and additions, it became clear that the human genome has genes similar to those in chimp, gorilla, orangutan, and even guinea pig and rat genomes," and thus, the once close chimp- human relationship had been severely shattered. Furthermore, the Y chromosome genetic comparison has proven that a chasm exists between humans and chimps.

"How did they rationalize this conclusion?" my wife Pam asked.

"The writers concluded that 'enormous vertical transmission of genes has occurred in human evolution via viruses and other vectors, building the human genome from scores of different animals. In addition, hundreds of genes found in humans have been found nowhere else and thus are the result of gene duplication.'

"Of course, to claim that the source of these genes is from somewhere else only begs the question, how did they get the genes in the first place? Second, the only evidence we have of gene transmission is the fact that some genes, or similar genes in various animals, are also found in humans," I told Pam.

The loss of my grants and graduate students caused not only depression, but severe migraines, vertigo, and other major health problems. On top of that, I arrived home from Cornell one Monday evening to find an empty house. Thinking that my wife, Pamela, and two children were out shopping, I cleaned up the kitchen and tried to relax and read the latest issue of *Nature*. At about 7:00 p.m. I received a phone call from my wife that would change my life forever. She was going to file for a divorce. She could no longer deal with the long-term stress in our life due to the evolution conflict, the court struggle, and the hate directed at me.

"Look," she told me, "why don't you just conform? Just accept evolution as a fact and do your research? This has ruined your life, has created enormous stress in my life, and in our marriage as well. It is not worth it!" she said, with stark anger in her voice.

"How was I to know it would come to this?" I replied. "I never in a million years expected well-educated professors would react so irrationally and dogmatically to someone questioning their Darwinian worldview that has been falsified over and over by scientific research."

"Well, now you know," she answered. "And it has cost you everything, and I cannot deal with the stress anymore. I am asking for a divorce."

"But you are a Darwin Doubter as I am," I added politely. "What happened to our vows, for better or worse?"

"Yes, I am, but I have never had any problems. I work with children and see miracles that only intelligent creation can explain. I probably have more faith than you do. I believe it is due to my faith, but you base your conclusions on scientific facts. But I have not testified in court, done any speaking, or written articles and books as you have. That's what got you into trouble."

She then added, before I could finish, "Look, I'm sorry … I am tired of the hatred, the nasty e-mails, and the hateful slanderous letters published in the newspaper about you. And, yes, the death threats as well. You should have thought about this before you went public. I kept my beliefs to myself and avoided all of these problems. I am sorry, but this has been enormously stressful for me and the kids, and we cannot, and will not, deal with it any longer. And as 'for better or worse,' there has been far too much 'worse' in our marriage."

"So, Pamela, you are abandoning me? Is that what you are saying?"

"I am just trying to keep my sanity and survive. That's all. I cannot take the death threats, the hatred, the fear, and the whole situation you are in. Our first years together were wonderful, but all that changed when you openly started to question Darwinism."

"What about our wedding vows, to love and cherish, in sickness and in health, in death only shall we part. Does that mean anything to you?"

"Times are different now. I just need space and time now."

She then hung up. I did not know where she was calling from, and she would not tell me. I was absolutely devastated. It was clearly the worst day of my life, bar none.

After she hung up, I thought about her conversation. I knew she was influenced by the faculty wives, many of whom were divorced, and, I later learned, they told her she needed to put herself first. If she was not happy with our situation, she should get the space she needed to seek out happiness and put herself first, as the women's lib groups she was involved in stressed. It was obvious that she was adversely influenced by them and their values.

To attempt to deal with this awful day, I drove to a supportive friend's house, a minister, David Heilbron, who was a Jewish convert to Christianity. After I told him what happened, he tried to help me put things in perspective. After we talked for over two hours, just before I left, he said, "Now that you realize Darwinism is false, what does that mean?"

Without waiting for me to answer, he answered for me, "If evolution is not our creator, what is the only other choice?"

"Do you mean God?" I answered. He then asked if I had ever read the New Testament. I said, "No, I have not." He added, "You know, Christ was a Jew."

I said, "No, no, he was a Christian!"

"Actually, he was Jewish, and so am I, but I am also a Christian. I am a Messianic Jew."

"OK. You got me there."

"How can you answer this question about Christ if you have not read the primary book about Him? Isn't that what evolutionists do? Ignore the evidence?"

"Yes, it is. I guess I'd better read the book first," I responded with a fake smile. He had helped me to feel a little better, I had to admit. Only a little though.

At the lowest point of my life, I began to read the New Testament. That began a series of questions I threw at David.

A week later, an "in the closet" ID supporter I knew came over to support me. In the end, he said to me, "Why don't you abandon your beliefs, or at least go into the closet. I know several Darwin Doubters who have done the same, and some have managed to resurrect their career. Some have even done well. I do not know if they still believe that Darwinism has been falsified, but at least they have all done well since they stopped opposing Darwinism."

"I cannot do that," I responded, somewhat surprised at the suggestion. "I would be living a lie."

"You do not have to lie; just tell people you have reevaluated your position and don't give them any details. Let me ask you, what is the effect of natural selection?"

"That is clear. It reduces the level of variety, the opposite of evolution. Assuming the peppered moth example was correct, at the start, we had light and dark moths. If the black ones were more easily seen by birds than the light moths, the black ones would be selected—eaten by birds—and the black ones would eventually go extinct."

"Correct. In the struggle for survival, some will lose, others will win. Thus, in the end, winnowing out the weak will result in less variety, not more, as evolutionists claim."

"That is well-documented," I answered.

"And mutations have what effect?" he asked.

"Most are near-neutral, thus will accumulate and will eventually cause degeneration of the genome, causing aging and death of the organism. It will also cause degeneration of the entire genome of the species so that the species will also age, and eventually genetic meltdown will occur, causing extinction of the species."

"So evolution is true, it is just going the wrong way, correct?"

"Yes, the research is very clear on that," I answered with confidence.

"So you are an evolutionist!"

"Yes, but not by the definition used by scientists and the public. Here is how the famous Oxford University militant Darwinist Richard Dawkins defines it: 'Evolution is selfish genes surviving.'"

"True, but your problem is the labeling, not so much what you believe. Accept the label and move on with your work and life."

"I think this is deceptive."

"Maybe, but what if a crazy man foaming at the mouth comes to your house and wants to know where your daughter is? Are you going to be fully honest with him and tell him that she is upstairs in her room?"

"Of course not. That would be foolish, to say the least."

"Well, have the Darwinists behaved rationally towards you about evolution?"

"No, not at all. They have behaved very irrationally towards me, have refused to read my writings, or even rationally consider my conclusions and ideas. One library put several of my books in their collection, but they were, in this case, shelved in the same area of the library as hate books, such as those supporting the KKK!"

"Well, why is this situation different from the case of the madman wanting to harm your daughter?"

"It is very different."

"Maybe in some ways, but not in others. These Dogmatic Darwinists are out of their minds, just like the madman!" he responded with anger.

"I agree. They are not thinking clearly, but in many ways, my situation is very different from a madman who wants to harm my daughter."

"Look, you have lost everything of importance in your life, your career, your reputation, your grants, your research, and now your family and health. What you are doing has not worked and will not work. They are Dogmatic Darwinists and cannot be reasoned with, correct?"

"You are very correct there," I nodded slightly in agreement.

"I think their hatred for those who reject Darwin is so great that they will do what they can to stop your research, even if it does end up curing some types of cancer. I am reminded of Forrest Mims, who edited a magazine called *Probe!* for amateur scientists. The magazine, the only one ever published devoted to this topic, got rave reviews from critics, but the Darwinist fanatics boycotted it and made sure that it failed. Their hatred was such that they would rather have no magazine for amateur scientists than one edited by a Darwin Doubter."

I thought about this option and decided that I would attempt it if only to prove the intolerance of Darwinists. I had been working on several important ideas for treating cancer and, if they worked out, I planned to write a book about my experience. After all, this was my goal ever since I started college. I was not a Christian then, and later I would regret my decision. I was then depressed, angry, and despondent. But let me finish the story.

I gave an interview to the local newspaper explaining that I had rethought my position and now accepted evolution. I was an evolutionist, I told them without defining the term. I noted the example of cat evolution, that from one cat type evolved all cats. I was not being disingenuous because even creationists accept this conclusion, which they call variation within the Genesis kind. The story hit the wire services, and within weeks, I did an interview for the *New York Times*, then *Discover* magazine, and even *Science* and *Nature*. Soon, I was on the lecture circuit and asked to tell my story of how I became a Darwinist. I often tactfully told them that I wanted to talk about my cancer research. I would briefly lecture on my "new" views, but I made it clear that I wanted to talk primarily about my cancer research, on which I usually spend most of my time discussing.

About this time, Professor Herzig invited me over to his house for dinner. Before we sat down to dinner—his wife was a great cook—he said the following to me:

> I want to apologize for how we treated you. I also want to admit I flirted with Intelligent Design while in college. My anatomy and physiology professor in my PhD program would often say things in class like "this system or organ is clearly well-designed." Later, I asked her in private if she believed it was really designed. She said she did, and that design was the only reasonable explanation for the human body and its anatomy, not Darwinism. She was an excellent instructor and really knew her anatomy, so I very much respected her opinion.

"That is most interesting," I commented, somewhat surprised, and wondered where this conversation was going.

He continued:

> Then I took a zoology class at the University of California, Berkeley, in 1968, with a then-unknown professor by the name of Richard Dawkins. He talked a lot about a theory he called selfish genes, meaning that our bodies exist only to serve our genes. The class was about zoology, but Dawkins spent most of the term lecturing about evolution and against creation. I was very interested in his ideas because I had accepted the design argument, so I thought I would talk to him after class about my concerns. Dawkins told me after class that he was once also persuaded by the argument from design, an argument for the existence of a god, or a creator, based on the evidence for order, purpose, and design in nature, and that he once had even embraced Christianity. He began having doubts about the existence of a god in his mid-teens when he learned about evolution in school. He then concluded that the theory of evolution was a better explanation for life's complexity and ceased believing in God.

"That is a very interesting story! I never knew this about you, or about Dawkins," I said, very surprised.

"Well, I have never told this story to anyone because I am ashamed that I once flirted with what is today called Intelligent Design. I later learned that my anatomy professor was terminated. I never found out why, but I am sure it had something to do with her rejection of Darwinism. After I took Dawkins's class proving evolution, I felt she should be fired, even though she was an excellent, very knowledgeable teacher."

"Very interesting!" I thought, amazed.

"But because she was such a good teacher, she did much harm because she caused students to have doubts about the unifying principle in science, namely, evolution. She caused me to doubt Darwinism, and I am sure she did the same to others."

After a great meal, I left for home, my home now without my family.

Pam soon married a man who was, to say the least, very different than me. She was stunningly attractive, so it did not surprise me that she remarried so quickly. Her new husband was a repairman, not involved in academia, and did not want me around. I tried through the court to exert my visitation rights, in the end, to no avail. My children were now totally out of my life. All they had to do was claim that I was an abusive father, and the court took away my rights. It was that easy!

My response to this deteriorating development was to bury myself in my work.

Chapter 20: I Am Now Part of the Club

In the meantime, other faculty began to warm up to me. Several even apologized for how poorly I was treated. All mentioned that they were glad I finally saw the light. The research committee at Cornell then supported obtaining new grants for me and assigned several graduate students. I soon had a three-million-dollar grant from the National Science Foundation and a new lab. In a matter of months, I had several research articles accepted in peer-reviewed journals, one in *Nature*. Life was surely wonderful as an evolutionist.

One case that was especially important and which helped to get me back into cancer research was that of a Mr. Fowler. After he was released from the service, he had planned to go to Sweden to study for a PhD at Uppsala University. That year, however, Fowler discovered an oddly shaped mole on the skin over his stomach that tests proved was cancer. By the time it was biopsied, the cancer had spread to his entire body. The doctors removed the skin where the melanoma had appeared, but the prognosis was grim. The doctor said that Mr. Fowler had, at most, only six months left to live. So, he went home and waited to die. When he felt the slightest twinge in his body, he wondered, "Is this the cancer I feel? Am I going to die soon?" But he didn't die, and after about a year, he decided that waiting to die was no way to live, so he decided to pursue his PhD anyway, which he completed in three years.

The doctors were so astounded by his complete remission that they sent Fowler to Memorial Sloan-Kettering hospital in New York to test his immune system to determine if that was the reason he beat his cancer. Although his immune system was weak when he first developed the cancer, it became remarkably effective once the cancer began to grow. The cancer itself triggered an aggressive immune response. They concluded that as a result, he was cured by his own immune system.

Ten years later, he was given a diagnosis of testicular cancer and underwent a painful procedure that involved injecting dye into his feet that circulated throughout his lymphatic system. An X-ray showed that the cancer had spread everywhere. He then lit up like a Christmas tree. Though they didn't tell him at the time, they never had a patient who had survived that kind of cancer. The doctors did extensive radiation treatments, and he still has tattoos all over his body put there to guide the radiation machine. Once again, the cancer disappeared due to his immune system's remarkably aggressive resurgence. I soon learned about more of these cases and felt that if we could just help the body do its job, the cancer would be cured without radiation or dangerous chemicals.

My research team once asked me for some basic details about my research, and I explained to them the following:

"I am looking at a way of enabling the body's immune system to aggressively act against cancer cells. I believe that, in harmony with Dr. Steven Rosenberg, the cure for cancer lies in the immune system. The immune system normally kills cancer cells; why does it not in all cases? As one ages, the immune system becomes weaker, and this is part of the problem. We know that all cells have over three hundred different kinds of receptors and other glycoproteins, lipoproteins, or lipo-glycoproteins on the cell's surface, such as folate receptors and CD4 and CCR5 receptors. Cell-surface receptors play a critical role in the biological systems of all animals. They make up 30 percent of all human proteins. I am also interested in the family of the nearly one thousand G protein-coupled receptors and the tyrosine kinase receptors. We also know that these receptors change in quantity and type according to the conditions inside the cell.

"Furthermore, when the cell is infected with a virus such as polio, this results in antigen display on the infected cell's surface. This triggers the immune system to destroy the infected cells and, consequently, the virus inside. The immune system is so effective that almost always, except in immunocompromised persons, every one of the millions, or even billions, of infected cells is rapidly destroyed. Thus, thanks to the immune system, almost everyone who gets the flu defeats it. What I am doing is looking for cell surface markers indicating that the cell has become a cancer cell. We know the function of some of these cell surface markers, and some reasons why immune therapy is very successful in some cases but fails in most others.

"We need to identify a universal cancer cell marker for each cell type, such as breast cells, that I am working on now, that will trigger an aggressive immune response. Or we need to identify the different markers that are unique to cancer cells, as there may be more than one. We have already identified several excellent candidates.

"We know the immune response varies from weak to so aggressive that anaphylactic shock results, and cytokines, such as interleukin 2 made by some T cells, induce immune cells to rapidly proliferate. So far, we have achieved promising results in my lab. Even if breast cancer cells spread to the liver and brain, which they often do over time, the cells are still breast cells, and thus the immune system will effectively destroy them. That, in summary, is my work, and I am very excited about the progress we are making.

"I realize this technique may not work, but because in many cases the existing treatments often fail, we have to explore other approaches. Due to the fact that cancer is the number one killer in the United States and much of the rest of the world, we must find a cure.

"As you know, I added, the current major cancer treatment is radiation or chemicals that disproportionately affect all fast-dividing cells, including not only cancer but also hair cells (thus one undergoing chemotherapy loses his or her hair) and gastrointestinal cells (thus nausea is common with chemotherapy). It works because cancer cells do not have the time to make the required repairs caused by the radiation or chemotherapy. This is because cancer cells do not rest as do non-cancerous cells, but as soon as cancer cells divide, they divide again, thus not allowing time to make the needed repairs. On top of this, cancer cells also typically have a damaged repair system. These treatments may prolong life, but if they miss a few cancer cells, the cancer will likely return. Only the immune system can effectively destroy all cancer cells."

Other researchers were on a similar path. Dr. David Porter of the University of Pennsylvania used genetically modified versions of the patient's own immune cells to seek and destroy their cancer cells, in this case, leukemia cells. Two of the three patients he treated for cancer were judged in remission within months. The goal was to modify the immune system to enable the body's own immune system to eliminate the cancer cells using the T cell type of leukocytes. T cells identify aberrant cells and pathogens, which unleashes the body's attack against them, involving cytokines and chemokines. They identified surface proteins, in this case CD19, that are found only on the surface of B cells. The therapy would kill all B cells, both normal and cancerous, thus a technique was required to compensate for the healthy B cell loss, such as stem cell transplants or immunoglobulin injections. Our goal was to eliminate only cancer cells by a similar technique.

After obtaining and publishing some very promising results, I was contacted by Pfizer, Inc., the world's largest research-based pharmaceutical company. They saw my work as promising and elected to pour millions of dollars into my research, and after several years, we began clinical trials. We found that several antigenic epitopes resulting from cancer progression responded positively to our clinical protocol. Thus, the specific protocol had to be tailored to each patient. Further research has allowed us to determine which relevant antigens would be found on each type of cancer, allowing us to tailor the treatment to each patient. Thus, we developed a set of specific antigens that produced the antibodies that primed the immune system to aggressively attack the cancer cells. Other labs were injecting polio cells into the glioblastoma type of brain cancer, with promising results based on the same idea.

I spent much of the next year in full-time research with a team of researchers, and the clinical trials looked extremely promising. Of the forty-nine patients, all in the terminal stage of breast cancer, forty-seven showed significant improvement. After a year of follow-up, forty-five patients were considered cancer-free. Soon, we expanded the trial to over two hundred patients, all in end-stage cancer. The ideas I had several years ago were working!

These results created a bombshell in the media, and several women's groups pushed for more rapid FDA approval. Our protocol was soon researched for other cancers. Within less than a year, our treatment system was successfully applied to pancreatic cancer, colon cancer, and liver cancer.

Chapter 21: Meeting Malinda

I had as many as 25 graduate students working for me at any one time, many from China. One, Malinda Chu, was one of my best students ever. She had no problem working in the lab at all hours, often after all of the other students had left. We worked well together and would often talk about our separate worlds—actually, most often she talked about her world, and I would listen. I rarely talked about my personal life with students. She was also the cutest thing I had ever seen. She had a touch of oriental and yet looked like an American model. She was born in China, but I wondered about her first name. Malinda was not exactly a common Chinese name, so I asked her about it.

"My father was an American priest, Gilbert Coulter; my mother, a young Chinese girl. They had an affair, and she became pregnant with me. Gilbert was removed as a priest and became a professor in a Catholic college in China but would not marry my mother. He was torn between his love for his church and my mother but supported her until she married my stepfather. My dad was a wonderful man. He was always there for me and treated me like a queen. My mother said he spoiled me rotten. I was anxious to please him and worked hard to do so. He also taught me English, helping me to become fluent in the English language."

The talk about Malinda and me eventually caused problems at the university. I finally confronted her with my concerns. She was almost thirty-three, and I was then thirty-nine. I had packed a lot of living into those thirty-nine years.

"You know that a lot of talk about us exists and could cause a problem. We spend a lot of time together, an awful lot," I told her with obvious concern written all over my face.

She answered, "Yes, I am very aware of that fact."

"You also know that this talk could cause me to lose my position?"

She answered, "Yes, I am very aware of that as well."

"And likely my career as well," I noted.

"Yes, I am aware of that possibility," she answered with a serious look on her face.

"I think you need to leave the lab when everyone else does, and we need to stop spending so much time together. It will be hard, as you are the best graduate student I have ever had. It will be difficult to do the work alone, but I do not want to risk losing my career again."

"Yes, I understand that, but there is another option," she replied.

"And what is that?" I asked.

"Look, we have worked very closely together for almost a year, closer than anyone in my life, ever. I know that you treat me in a very special way, like no other student," she added.

"And," I asked?

"I also know that you are divorced."

"Yes, that is true, I am."

"Thus, you are free to marry."

I did not need any more hints. She was untouchable as a student, but I saw her as a dream come true. Not since Pam have I had any feelings for someone as I did for her.

She took my hand and said, "Well, in America, the man is supposed to ask the girl."

So, I did, and asked her, "Will you marry me?"

"Yes, yes, I will!" she answered, with joy written all over her face and a smile as I have never seen on any woman, ever.

"So, we are engaged?" I commented. I felt like a high school kid, shy yet excited.

"Yes, we are engaged!"

We embraced and talked about a wedding date.

The next morning, we announced our engagement to the entire lab. Most of the staff were not in the least bit surprised. The age old, "I was the last to know," thought struck me then.

A day later, my dean called us into his office to talk, one at a time, and grilled her. "Was any coercion involved? Any promises of raises or a promotion?" he asked.

Malinda answered: "Well, I dropped a few hints, and I will be thirty-four in less than two months, so I think I can make these decisions on my own. After all, Marie Curie married her professor, and they worked very well as a team."

In the end, he answered, "You are right. I have no power to interfere. And I was young once, so all I can say is sincere congratulations."

We were married shortly after she turned thirty-four. As an evolutionist, I got my career back and even got a wonderful wife. Life was now very good for me, very good.

Malinda had become a Christian in China and was also a creationist, so we had no conflicts with the Darwin issue. I was a non-practicing Jew, so I had no problem with her faith and thought it could be a plus. We would have many talks on religion. In this area, she was clearly way ahead of me, and I respected her views. I also began to grow beyond my nominal Judaism.

With my new wife by my side every step of the way, we, and several other researchers, fully documented our findings. I was again regarded as a well-respected researcher. My Intelligent Design days were well behind me. The fact is, in a few years, I was awarded the Nobel Prize for my cancer research. It was then that I retired from Cornell and decided to write this book. This would be my story, showing what refusal to bow down to the dogma of Darwinism can cost. Yes, I was an evolutionist, the theory was true but going the wrong way.

The same intolerance almost immediately surfaced when word got out about my turncoat heresy.

When I detailed my beliefs, I was condemned again, this time also as a turncoat or worse, and I actually had several awards rescinded; the scientific world had egg on its face, as they should. I felt that my book, in the end, could be much more important than my cancer research and even my Nobel. The intolerance in science must stop. Scientists must take off their blinders and let the evidence, and only the evidence, speak. Towards this end, I wrote this book.

I had a difficult time finding a publisher, and when the book finally came out, Cornell, my own university, refused to put a donated copy into their library, as did many other colleges. Some things never change. I was also told that my admission to accepting creation again was a tragic ending to a once brilliant career. Several people informed me that I was a great scientist but tragically had lost all my reasoning ability in my last few years. The intolerant science establishment could not do much to damage me now, and although my turncoat reputation hurt me greatly, both psychologically and financially, I felt that I must follow the evidence wherever it led. As Martin Luther once said long ago, "I can do no other."

Epilogue

After a few years, I realized that I harbored a great deal of guilt over my deception but was helped to deal with it by my Messianic Jewish friend. He stressed his belief that the acceptance of Christ can wipe away even the most egregious of sin, and I knew full well that I was a sinner, and my deception helped me more than anything to realize this fact. But that is another story and another book.

Appendix

The Unethical Tactics of the Evolution Attack Dogs: P.Z. Myers and Attitudes of Evolutionists in Response to the Bergman-Gutsick Gibbon Debate

By Jerry Bergman, PhD

What this appendix is about: In response to valid concerns about the origin of life and the claims by evolutionists about evolution, instead of a rational response, evolutionists often resort to mocking the questioner and name calling. Instead of mentioning the Urey-Miller experiment performed by Harold Urey and Stanley Miller in 1952, as often done in the text books, evolutionists resort to calling the questioner stupid, a "creationist liar," and an "imbecile moron creationist." This response hardly helps the evolutionists, rather it hurts them greatly.

I recently had a debate titled *Does the Human Fossil Record Support Human Evolution*? The debate analyzed the claims of evolutionists that the fossils support the theory that humans evolved from a common ancestor of humans and chimps. My debate opponent was the enormously popular Erika "Gutsick Gibbon" (pseudonym), a Ph.D. student in paleontology. Erika was a young-Earth creationist educated at a Christian school from kindergarten to eighth grade. She then went to a secular high school where she was exposed to evolution in her textbooks, causing her to reject her creation belief.[60] Many aggressive evolutionists were once creationists.[61] This partly explains her strong interest in the issue. It also supports the common poor education about this issue in many, if not most, Christian schools. Erika related that her parents—her father was a medical doctor—were some form of creationists.

So far, the Gutsick v. Bergman debate has had over 70,000 views on YouTube. Erika was in every way polite, respectful, and considerate in presenting her position, which was well thought out and well-illustrated. She surprised me with how well read she was on many topics. The host, Donny

---

[60] Godless Engineer (2020 Jan 03) *How Gutsick Gibbon Erika Left Young Earth Creationism*, https://www.youtube.com/watch?v=4xTea7d5dtY

[61] Bergman J (2019) *Darwinism is the Doorway to Atheism*, Leafcutter Press, Southworth, WA.

Budinsky, it appeared to me, was a little hard on me and, in contrast, very accommodating of Erika.

Erika's main argument was to describe several fossils that she claimed gave evidence of primates that could walk upright, either occasionally or as their main mode of traveling. Her argument was that these examples were evidence of evolution from the quadrupedal mode, as a chimp uses, to the bipedal mode of transportation used by humans. My response was that I had no problem with some nonhuman primate that could walk upright. I then discussed the problem of genetically converting quadrupedal to bipedal locomotion by chance evolution.

In short, she believed on faith that the fossils she presented proved human evolution, where I felt her evidence only showed that some primates could walk upright, something very different than proving chimps and humans have a common ancestor that looked something like a chimp. I have authored a large book that documents, by citing over several thousand references to the professional paleontological literature, the problems with the human fossil record.[62] Thus, I felt to repeat the Dr. So-and-So made some claim against one interpretation of a claimed prehuman fossil and then quote another paleontologist who countered his claim would not result in a productive debate. Therefore, I decided to document the fact that none of the claimed missing links are actually evidence of human evolution because evolution has no viable mechanism to cause macroevolutionary change.

I first showed that the evidence used to accept human evolution for the first century since the publication of the *Origin of Species* in 1859 has all been decisively refuted by science, yet was still accepted as solid proof that hydrogen-to-human evolution was an unequivocal fact. Next, I documented that the classical view of human evolution, called *The Progression,* from some common ancestor to modern man was rejected, and in its place was a set of different primates and not a progression leading up to humans (Fig. 1, Fig. 2).

---

[62] Bergman J, Line P, Tomkins J, Biddle D (2021) *Apes as Ancestors: Examining the Claims About Human Evolution*, Bartlett Publishing, Tulsa, OK.

Figure 1. Erika was supporting the now-rejected, pro-evolution "Progression" icon shown here.

Figure 2. This is another example of the now-rejected, pro-evolution "Progression" icon.

Lastly, I documented that no viable genetic mechanism exists for evolution from molecules to man, a major problem that Darwin had and that still exists today. The mechanism claimed today, damage to the genome, called mutations, is the primary source of the new genetic variation that natural selection selects from to cause evolution from prokaryotes to humans.[63] Erika was unable to refute any of the information I presented; thus, her fossil discussion was not supportive of so-called missing links from our theoretical common ancestor to modern man.

Figure 3 illustrates the new arrangement of the fossils that are purported to support human evolution. This new family tree was based on hundreds of fossils uncovered during the last five decades. It places the once claimed human ancestors in four distinct species, namely Homo, Paranthropus,

---

[63] Bergman J (2019) *Science is the Doorway to Creation: Nobel Laureates and Other Eminent Scientists Who Reject Orthodox Darwinism*, ed. Kevin Wirth, Leafcutter Press, Southworth, WA.

Australopithecus, and Ardipithecus with no claims of evolution from one species to another.[64,65]

Figure 3. The current belief. The "Progression" has now been discarded by most paleontologists and replaced with a distinct set of primates that fall into four different groups.

Comments About My Debate Performance

In the 4.5-hour debate, I likely made at least one mistake in my presentation and was curious to determine if someone else caught any actual errors. I was also very interested in the responses to my presentation, which had over 70,000 views.

The 1,884 comments posted in the comment section of the YouTube debate strongly favored her and often mocked my presentation. It is clear that most were committed evolutionists who did not understand or comprehend my PowerPoint presentation. If in a debate an evolutionist lacks evidence, in my experience they commonly resort to name calling. In this case most of the comments were name calling, which is a common response for Darwinists

---

[64] Bergman J (2009) The Ape-to-human progression: The most common icon is a fraud, *J Creation* 23(3):16–20.

[65] Bergman J (2024 Dec 04) Time to unlearn the evolutionary march of human progress. New discovery adds more evidence against the human evolution progression. *Creation Evolution Headlines* https://crev.info/2024/12/time-to-unlearn-the-evolutionary-march-of-human-progress/

and does not give their views credibility. I authored an entire chapter covering the common response of evolutionists. Instead of mustering solid arguments, they resort to ridicule and name calling.[66] Some typical examples from the YouTube comments section are presented here, uncorrected for spelling, punctuation, and grammar:

- You can be certain that he will go back to his echo chamber and claim victory. They are exactly like flat earthers.

- This was an absolute beat down. Basically 100 gibbons vs one Neanderthal

- I want to say as someone who grew up within a young earth creationist household, I accepted evolution a decade ago, and this debate actually gave me so much good information I had been unaware of and am even more convinced thanks to the evidence presented by you Erika. And the great part is learning this was fun! It is always frustrating when people are presented with information and dig in their heels when they could otherwise enjoy the facts as they come and be excited by learning

Examples of positive comments:

- Dr. Bergman acknowledged the existence of these old bones, but Erika never demonstrated HOW they could be construed as "Human Transitions." She never stated what tools and social evidence was found along with the bones. Being "Human" is more than anatomical; it's behavioral as well.

- Dr. Bergman clearly asked her what mechanism she had to demonstrate that these were "Transitional." Just pointing at similar features doesn't cut it. You need a mechanism to make the major changes occur.

Another common example of the dominance of derogatory comments by supporters of evolution is Erika's debate with Kent Hovind titled "Epic Debate: Gutsick Gibbon Vs Kent Hovind | Human Evolution | Podcast."[67] The debate video had 234 views and 11,240 comments, the vast majority of

---

⁶⁶ Bergman J (2024) Chapter 2: The name calling problem, *Silencing the Darwin Skeptics,* second edition. Leafcutter Press, Southworth, WA, 42–78

⁶⁷ Modern-Day Debate, *Epic Debate: Gutsick Gibbon Vs Kent Hovind | Human Evolution | Podcast,* https://www.youtube.com/watch?v =VXFojsRxJLg

which were negative about Kent Hovind, such as the following (again, with spelling, punctuation, and grammar uncorrected):

- Kent Hovind has gone from being a man who spoke passionately about what he believed in to a sad old grouch debating whoever he can find in a desperate attempt to stay relevant. Well done Erika for kicking his ass, just like everyone else has

- Gibbon's opening statement was so well-thought out, so descriptive, and with his third sentence, you can tell Hovind didn't listen to a darn syllable of it.

- Creationists never win debates, but this one is such a spectacular loss haha. Erika is fantastic.

- I am entertained by the imbecil creationist

I also looked for reviews of my debate with Erika on the internet and found several. None of the reviews I read made any claims of errors in my presentation. By far the most common response was mocking the creation worldview and me as well. The opposition took several forms. Some comments were respectful but were poor arguments. Some do not address the issue but generally denigrate young-Earth creationists. Some were appallingly nasty. One review that stands out was not a review but lambasted the creationists in general and my performance specifically in the debate with Paul Myers, a name he does not use, preferring to go by P.Z. Myers instead. He is an associate professor of biology at the University of Minnesota–Morris. I also, many years ago, had debated him in a live debate in Minnesota. The following is a summary of my debate with Erika and my responses to his "review."

P.Z. Myers's Review of the Gutsick Gibbons vs. Bergman Debate

His "review of the debate," indented below, was predictably nasty, consisting mostly of name-calling, which was a stark contrast to the performance of Erika.[68]

Bergman, as I predicted, was a sloppy mess with a scattershot collection of slides that were mostly off-topic and irrelevant, and was [sic] full of wrong

---

[68] Myers PZ (2025 May 18) Stop letting creationists host your ideas, Pharyngula: Evolution, development, and random biological ejaculations from a godless liberal, https://freethoughtblogs.com/pharyngula/2025/05/18/stop-letting-creationists-host-your-ideas/

examples that didn't make his case. Would you believe he talked about Nebraska Man, a hoary old chestnut of tabloid excess that never had the support of the scientific community, presented alongside Piltdown Man as evidence that the fossil record was fake?

I mentioned the major claims used by evolutionists to prove their worldview as fact, all of which have been credibly rejected today. My point was that these examples were touted as irrefutable evidence for hydrogen-to-human evolution. As a result of this now-discredited evidence, evolution as fact was taught in the textbooks and used in court cases, including the Scopes Trial, as irrefutable proof. My point was, as was true for the last century, the evidence for evolution likewise will continue to be demolished, and many scientists will continue to believe fervently in evolution as defined above. Myers totally distorted my testimony.

> How about the claim that Australopithecines were just the bones of pygmies?

He implied that I claimed this, which I clearly did not. I only mentioned that some authorities took this position. In the quotes below, Myers's name-calling is highlighted in italics:

> But I almost gave up in the first few minutes, before either had a chance to speak, when the screen loaded and there across the top in big capital letters was the banner "STANDING FOR TRUTH BIBLICAL MINISTRIES" with the logo for that disgraceful organization popping up throughout. The moderator/host was that *smug twit*, Donny Budinsky, a hardcore young-Earth creationist *with no education in science, geology, paleontology, or evolutionary biology*, who promotes these *inane* "debates" between creationists and sane evolutionary biologists. WHY? This was a promotional event for the *dumbest collection of ignorant yahoos on YouTube. These are terrible people*, and yet so many science educators will voluntarily send traffic their way, and, by the way, platform *dogmatic buffoons like Jerry Bergman*.
>
> ...We don't need *grifting yahoos like Donny Budinsky* to organize and host these "debates," and if you *ditch mind-numbing parasites like Bergman*, you don't even have to waste time on them — Erika had a robust, informative 45 minutes of science talk *imbedded in the superfluous, distracting garbage of the Jerry and Donny Show*, with an ad for creationism layered on top.
>
> ...I despise debates, but even worse are debates that donate unwarranted attention and respect to *lying apologists for anti-scientific claims*. Stop it, everyone."

My Response to Meyers's Rant

My main comment is that I would never expect such ranting from an adult college professor. He responded to none of the material I included in my presentation. His behavior is more like what I would expect from a poorly educated lower-class adolescent.

Select comments from readers of Myers's website below are taken from his website;[68] frequent misspellings were corrected; name-calling is highlighted in italics:

- …her part in a video, …. sounds interesting, but I don't want to see those *liars*

- …regarding debating *Xian charlatans* – it has always been thus debating *Muslim fundie charlatans*, as they only accept a "debate" if the host / moderator is one of them. It is not the ordinary Muslims (or Xians) I despise, it is the dishonest propagandists. May the Fenris Wolf take them.

- Alternatively, you can turn the creationists' garbage against them.

- This is why you never debate these people (there is *no intellectual difference between fascists and creationists*).

- The purpose is fundraising and in-group social reinforcement for the creationists. It has nothing to do with "debate." It's just pro-wrestling and merch sales, with science cast as the heel.

- And on the specific subject of the "debate," the fact remains that the fossil record, skeletal remains, genomic analysis, and other lines of evidence necessarily leave unanswered questions about hominin evolution but completely and utterly contradict YEC.

- I long ago refused to engage with creationists. …all we get are begets and sins and centuries of blood-stained history. Nothing at all useful. You'd think that if their god really loved us, he would've salted in some clues as to how the world actually works.

- I'm with Akira and Thought slime. …A "debate" is not something you can have with *fascists or creationists*. Their *worldview is so twisted* that there's no way you are going to change their minds.

- I just don't have the patience to *argue with idiots*.

- Indeed, and these ridiculous events are the only chance a scientist has to reach that audience. Such is the bubble that Christians have built around themselves. Mockery and tearing apart the foundations of their creed is far more effective.

- As for effectiveness, I can think of nothing that shuts down humans faster than making fun of them. Erika … probably read all of these hot takes before in the aftermath of the Bill Nye - Creation Museum debate. …she isn't as famous as Bill Nye, so she isn't propping up their **idiocy** with a chance to declare victory on a celebrity.

- I just don't have the patience to argue with *idiots*.

- We've been arguing with *idiots* for decades now.

- And here we …living in a failing democracy, currently being run by *Fascists* in the service of the *Oligarchies*. While being supported by almost half the population, the MAGAs and their associates.

- I'm all out of patience with the *idiots* these days, too!!!

- The state I live in has decided to hand out money meant for public schools to nonpublic schools.

- … trying to talk over each other with *meaningless emotional appeals*. Internet debates (and any discussion with *Xian terrorists*) are like trying to reason with a rattlesnake.

- How do you debate someone who is allowed to use "it was magic" as a legitimate explanation in their favor?

- At a certain level of *ignorance*, one gets impervious to any kind of reasoning.

- The problem is, a robust percentage of the population is now capable of believing absolutely anything, and their hallucinations are getting wilder and wilder as time passes. Apart from their self-harming behaviors (no-vaxx, weird diets, bleach gargling, what have you), they are easily manipulated by whoever finds it convenient to do so. A new generation of useful *idiots*, if you will.

- … knuckle-dragging mouth-breather *Xian terrorist maggots* I find that: 'logic eludes them.'

- I have seen multiple comments here and elsewhere along the lines of "we have to deal with such *stupid/ignorant/irrational* people now". …When were humans not faced with dealing with other humans who were *stupid, ignorant, or irrational*? The *dumb ideas* need to be stomped on with gusto.

- Seriously, what would you do in case like that, especially if that guy is anywhere near a sharp, metal tool?

From *Rate My Professors*

Curious about his role as a professor, I looked up the online comments on *Rate My Professor*. Select comments from his students include that he indoctrinates his students in evolution and against the creationism worldview. As is common, his student reviews were mixed but reveal that he manifested what can only be described as unmitigated hate for anyone who has the audacity to question his worldview.[69]

The following are unedited comments taken from the *Rate My Professors* website regarding Paul Myers as a professor,[69] with emphasis added in italics.

- I fell asleep in almost every lecture but easily got an A by looking back at the note slides and the textbook

- His connection with his students is very poor.

- Would never take a class with this man again after this experience. For a "fun change in pace," he assigned a five-question exam all essay. He expected each question to be answered in no less than two pages. The questions he asked were more suited for an ethics class and the majority of them were offensive. Would not recommend.

- Prof Myers is *heavily engaged in anti-creationism* which is great, but he is apparently 'too important' to substantively engage with students even in office hours. Luckily the material isn't too challenging but he seems lazy about writing tests and informing us what's in them.

- *PZ Myers, very famous...has made numerous appearances everywhere, especially on the evolution vs creationism* thing

- Interesting character. *Very into anti-creationism*. He can teach the course so that you want to learn. Not too hard, not too easy. One thing is he is always busy, always. Very hard to find him for office hours. Otherwise, *if you are a liberal person, you will like this prof. If you are conservative, you will want to strangle him in the first week.*

---

[69] *Rate My Professors,* Paul Myers,
https://www.ratemyprofessors.com/professor/123513

- I learned some useful concepts from the readings, but *the professor is arrogant and preachy. He dismisses views he disagrees with instead of trying to understand and encourage reasonable disagreement.*

- This guy sucked. PZ is full of himself because he is mildly famous, but he's lost his touch. He is a poor lecturer, *he is sarcastic and demeaning*, and rather than teaching better, he just curves grades.

- He locks himself in his office and won't answer his students knocks. His grading is INCREDIBLY inconsistent!

- I hope to never have this professor again. The scope of the class was to have 4 quizzes and 4 tests. I liked that idea, the quizzes being the benchmarks for how well you're doing and will help on the test. However, there were no quizzes, just tests for 100% of the grade. Professor is lazy and has terrible tests. Take someone else.

- *He's very opinionated.*

- Wasn't organized. The syllabus was from the previous semester, and he made significant changes, making it hard to know what was going on. This class was the most random class ever. The tests were on things barely covered or not covered in class.

- P.Z. Myers is awesome! He's really famous so it's like having a celebrity in class! *He's up there on the news with* [atheists] *Richard Dawkins and Jerry Coyne.*

- Alright this guy is sweet as hell, frank as hell and very good with words. He runs a popular blog and has been featured on several documentaries and gives talks everywhere.

My Conclusions About Myers Teaching Religion and The Debate with Erika

It is obvious that Myers teaches religion in his class. Specifically, he teaches against the creation worldview, which is legal; but the courts have ruled in over 200 cases that it is illegal to present information *in favor* of creation. In other words, as a result, he teaches students *what* to think, not *how* to think, which is what should be taught. This is clearly indoctrination which is similar to the Nazi policy of indoctrinating students that Jews are bad people trying to take over the world and must be stopped to save the Aryan race.

The debate with Erika also argues for detailed discussion in Christian schools of why evolution, as defined in this paper, did not, and could not, occur. Her evidence is controversial but not unequivocal. Erika has recently moved from being a theistic evolutionist to being an agnostic.

The other upload of the debate I had with Erika is here:
https://www.youtube.com/watch?v=yyoioiM7Dql&t=4404s

This one has more positive comments from creationists. Erika's subscriber base is 99% evolutionist, so those comments tend to be negative towards the creation worldview.

Other Books by Jerry Bergman

Evaluation of an Experimental Program Designed to Reduce Recidivism Among Second Felony Criminal Offenders. Ann Arbor, MI: University Microfilms Publishers, 1976. 886 pages

Instructor's Guide: Experiences in Math for Young Children. (Co-authored with Drs Deanna Radeloff and Rosalind Charlesworth). New York. NY: DelMar Publishing Co., 1978. 69 pages

Teaching About Creation/Evolution Controversy. Bloomington, IN, Phi Delta Kappa Educational Foundation, October 1979. Published as a fastback. Number 134. 45 pages. WorldCat lists 355 libraries with a copy.

Religious Objections to Blood Transfusions: A History and Evaluation. Los Angeles, CA: CARIS Press, 1979. 22 pages

"Peer Evaluation of University Faculty." *College Student Journal Monograph,* 14(3), Part 2, Fall, 1980. Chula Vista, CA. 22 pages

Understanding Educational Measurement and Evaluation. Boston, MA: Houghton Mifflin Co., 1981. 304 pages. WorldCat lists 196 libraries with a copy.

Jehovah's Witnesses and Kindred Groups: A Historical Compendium and Bibliography. New York, NY: Garland Publishing, Inc., 1983. 412 pages. WorldCat lists 506 libraries with a copy.

The Criterion; Religious Discrimination in America. Richfield, MN: Onesimus Publishing Co., 1984. 116 pages. WorldCat lists 43 libraries with a copy.

Community and social control in a chiliastic religious sect : a participant observation study. MA Thesis Bowling Green State University. 1986. 265 pages.

Jehovah's Witnesses, Vol. I; *The Early Writings of J.F. Rutherford.* (Ed. and Introduction), New York, NY: Garland Publishing, 1990. In 100 WorldCat libraries

Jehovah's Witnesses, Vol. II; *Controversial and Polemical Pamphlets.* (Ed. and Introduction), New York, NY: Garland Publishing, 1990. In 96 WorldCat libraries.

"Vestigial Organs" Are Fully Functional: A History and Evaluation of the Vestigial Organ Origins Concept. (With George Howe, Ph. D., Forward by David Menton, Ph. D. Washington University School of Medicine; Preface by V. Wright M.D., F.R.C.P., Professor University of Leeds). Terre Haute, IN; Creation Research Society Books, 1990, 97 pages. WorldCat lists 117 libraries with a copy.

Zur seelischen Gesundheit von Zeugen Jehovas. Germany: Druck Bielefeld. 80 pages, 1991.

A History and Evaluation of Noninvasive Medical Diagnostic Treatment and Research Techniques." Ann Arbor, University Microfilms, 1992, 498 pages

Jehovah's Witnesses and the Problem of Mental Illness. Introduction by Dr. Carl Thornton, Professor of Psychology, General Motors Institute, Flint, MI. Clayton, CA: Witnesses, Inc., 1992, 342 pages

Jehovas Zeugen und das Problem der seelischen. München: Claudius Verlag, 1994, 84 pages. Translated by Dr. Helmut Lasarcyk.

Blood Transfusions: A History and Evaluation of the Religious, Biblical and Medical Objections. Introduction by Dr. Robert Finnery. Clayton, CA: Witness Inc., 1994, 208 pages. Cover by Artist David Merrick.

Bluttransfusion. München: Claudius Verlag, 1995, 84 pages

Creativity: A Correlational Study Between Creativity, Teacher, Grades, Intelligence, Teachers Creativity Ratings, School Citizenship, Test Anxiety, and Copying Style. Master's Thesis. Wayne State University, Detroit Michigan. 178 pages. 1973.

I Testimoni Di Geova E La Salute Mentale. Roma: Editzioni Dehoniane 1996 Prefazione by Dr. Maurizo Antonello, Psicologo.

Jehovah's Witnesses: *A Comprehensive and Selectively Annotated Bibliography.* New York: Greenwood Press 1999. 352 pages. Volume 48 in the Greenwood Series of Bibliographies and Indexes. WorldCat lists 294 libraries with a copy.

Tumor Markers in Cancer Treatment. Master's of Science in Biomedical Science Dissertation. Medical College of Ohio, Toledo, Ohio 1999. 355 pages

Ward's Science Lab Safety Manual. Rochester, NY: Ward's Natural Science Est., Inc. 2002, 75 pages

Evaluation of Optical Radiation in Medical Diagnostics and Treatment. Masters Thesis. Medical College of Ohio, Toledo, Ohio. 2004. 78 pages

Persuaded by the Evidence (editor). Green Forest, AR: Master Books. 2008. 288 pages. with Doug Sharp. WorldCat lists 74 libraries with a copy.

Hitler and the Nazi Darwinian Worldview: How the Nazi Eugenic Crusade for a Superior Race Caused the Greatest Holocaust in World History. Joshua Press, Kitchener, Ontario, Canada. 2012.

C.S. Lewis: Anti-Darwinist: A Careful Examination of the Development of His Views on Darwinism. Wipf & Stock Publishers, Eugene Oregon. 2017

Silencing the Darwin Skeptics: The War Against Theists. Leafcutter Press, Southworth, WA, 2018.

Fossil Forensics: Separating Fact from Fantasy in Paleontology. Bartlett Publishing, Tulsa, OK. 2019.

Apes as Ancestors: Examining the Claims About Human Evolution. Bartlett Publishing, Tulsa, OK. 2020.

How Darwinism Corrodes Morality: Darwinism, Immorality, Abortion, and the Sexual Revolution. Joshua Press, Kitchener, Ontario, Canada. 2020.

Darwinian Eugenics and the Holocaust: American Industrial Involvement. Involgo Press, Peterborough, Ontario, Canada. 2021.

The Last Pillars of Evolution Falsified: Further Evidence Proving Darwinian Evolution Wrong. WestBow Division of Thomas Nelson and Zondervan, Bloomington, IN. 2022.

The Three Pillars of Evolution Demolished: Why Darwin Was Wrong. WestBow Division of Thomas Nelson and Zondervan, Bloomington, IN. 2022.

C.S. *Lewis' War Against Scientism and Naturalism.* Cantaro Press, Ontario, Canada. 2023.

Darwin's Replacement: Summary Edition. Lifetime Reference Guides, Inc., West Vancouver, British Columbia, Canada. 2023.

The Dark Side of Darwin. Green Forest, AR: New Leaf Press. 2011. WorldCat lists it in 237 libraries so far. Revised edition published in 2015.

Hitler and the Nazi Darwinian Worldview: How the Nazi Eugenic Crusade for a Superior Race Caused the Greatest Holocaust in World History. 2012. Kitchener, Ontario, Canada: Joshua Press. 356 pages. WorldCat lists it in 106 libraries.

Transformed by the Evidence edited with Doug Sharp. 2014. Southworth, WA: Leafcutter Press. 240 pages. WorldCat lists it in 19 libraries.

The Darwin Effect. Its influence on Nazism, Eugenics, Racism, Communism, Capitalism & Sexism. 2014.

Green Forest, AR: Master Books. 360 pages. In 247 WorldCat libraries.

Debunking Human Evolution Taught in Our Public Schools: A Guidebook for Christian Students, Parents, and Pastors. with Dr. Daniel A. Biddle and David A. Bisbee. Genesis Apologetics Press. Folsom, CA. second edition. 2016. 180 pages. I wrote chapter, 4, What about the Problem of Overdesign for Darwinism pp. 116–124; and chapter 8 Vestigial Structures in Humans and Animals pp 125–138.

C. S. Lewis: Anti-Darwinist: A Careful Examination of the Development of His Views on Darwinism. Eugene Oregon: Wipf & Stock Publishers. 160 pages. 2016. Listed in 322 major college libraries.

How Darwinism Corrodes Morality: Darwinism, Immorality, Abortion and the Sexual Revolution. 2017. Kitchener, Ontario, Canada: Joshua Press. 312 pages. In 14 WorldCat libraries.

Evolution's Blunders, Frauds and Forgeries. Atlanta, GA: CMI Publishing. 2017. 320 pages. So far in 10 libraries in WorldCat

Slaughter of the Dissidents: The Shocking Truth About Killing the Careers of Darwin Doubters. 2008. Revised version 2012. Southworth, WA: Leafcutter Press. 482 pages. WorldCat lists 72 libraries with a copy. Third edition published in 2014

Silencing the Darwin Skeptics. Southworth, WA: Leafcutter Press. 2016. 396 pages. Revised edition 2023.

Censoring the Darwin Skeptics. How Belief in Evolution is Enforced by Eliminating Dissidents. Southworth, WA: Leafcutter Press. 2024. 566 pages. Revised third edition.

Darwinism is the Doorway to Atheism. Southworth WA: Leafcutter Press. 2019.

Science is the Doorway to Creation: Nobel Laureates and other Eminent Scientists who reject Orthodox Darwinism. 2019. Southworth, WA: Leafcutter Press. 404 pages

Fossil Forensics: Separating Fact from Fantasy in Paleontology. 2017. Tulsa, Oklahoma: Bartlett Publishing. Revised edition 356 pages. So far in 22 WorldCat libraries. Revised edition Creation Summit. 2025

Useless Organs: The Rise and Fall of a Central Claim of Evolution. Tulsa, Oklahoma: Bartlett Publishing. revised version 2024. Revised edition 332 pages. 2 libraries

Poor Design. An Invalid Argument Against Intelligent Design. Tulsa, Oklahoma: Bartlett Publishing. 2024. 230 pages. Revised edition

God in President Eisenhower's Life, Military Career, and Presidency. Eugene Oregon: Wipf & Stock Publishers. 206 pages. 2019. In 67 WorldCat libraries.

The Methodist Darwin Syndrome: Consequences of Adopting Darwinian Theology. Lansing, MI:Looking Glass River Publishing. 2021. 278 pages.

Darwinian Eugenics and The Holocaust: American Industrial Involvement. Involgo Press. Peterborough, Ontario, CANADA. K9J 6X8 2020, 200 pages. www.involgo.com.

Darwin's Replacement: Summary Edition. West Vancouver, B. C. Canada. Lifetime Reference Guides, Inc. 23 pages. With Dr. Graham McLennan PhD and Thomas Rogers. 2021

How Great Evil Birthed Great Good. The True Story of Two Families. WestBow division Thomas Nelson and Zondervan. 2021. 180 pages.

Apes as Ancestors: Examining the Claims About Human Evolution. Tulsa, OK: Bartlett Publishing. Co-authored with Peter Line, PhD and Jeff Tomkins PhD . 2021.

Wonderful and Bizarre Life Forms in Creation. Edmonton, Alberta. Creation Science Association of Alberta, Canada. 2020. 144 pages. 2020.

How we are Amazingly Made. Lifetime Reference Guides. West Vancouver, BC Canada 2022. 40 pages. With Dr. Graham McLennan and Thomas Rogers. 2022.

Three Pillars of Evolution Demolished. Why Darwin was Wrong. 2022. WestBow Division of Thomas Nelson and Zondervan. 358 pages. In 5 libraries. 2022.

The Other Side of the Scopes Monkey Trial. At its Heart the Trial was About Racism. Wipf & Stock Publishers, Eugene Oregon. 2023. In 79 libraries.

Why Did God Create Viruses, Bacteria, and Other Pathogens. 2023. Bloomington, IN WestBow Division of Thomas Nelson and Zondervan. 162 pages.

Trials of a Darwin Doubter. 2023. Lansing, MI: Red Cedar Publishing.

Evolution's Dangerous Ideas: Eugenics, Lobotomies, Using X-Rays to Speed Up Evolution, and Other Dangerous Ideas Inspired by Darwinism. Cantaro Publications Jordon Station, ON. Canada 2024.

C. S. Lewis's War Against Scientism and Naturalism. Cantaro Publications Jordon Station, ON. Canada. 2024

Tackling Tough Issues. 2026 Creation Summit Publishers Lindhurst, Il. 2026. 216 pages.

Student Guide for Debunking Evolution, Edited by Daniel Biddle. Genesis Apologetics. 2016. 100 pages

Awakening of a Jehovah's Witness: Escape from the Watchtower Society. Hardcover Prometheus; Reprint edition (April 1, 2002) 328 pp 164 ratings on Amazon. by Diane Wilson (Author), Jerry Bergman (Author) Diane Wilson was the main author, Bergman the second author.

How Germany and the United States Produced the Holocaust. Kitchener. Not out yet.

Gods Biology. Darwin's Replacement. 2024. With Dr. Graham McLennan and Thomas Rogers. Lifetime Reference Guides. West Vancouver BC Canada.

The Darwinian Genocide in Africa: Darwinian Influence in the Development of Supremacist and Segregation Policies in Africa, Eugene Oregon: Wipf & Stock Publishers. 2026.